Vegetative Compatibility Responses
in Plants

Vegetative Compatibility Responses

in Plants

Edited by Randy Moore

LIST OF CONTRIBUTORS

J. R. Aist, Department of Plant Pathology, Cornell University, Ithaca, New York, U.S.A.

A. Bacic, Plant Cell Biology Research Centre, School of Botany, University of Melbourne, Parkville, Victoria 3052, Australia

A. A. Bell, United States Department of Agriculture, National Cotton Pathology Research Laboratory, P.O. Drawer JF, College Station, Texas, U.S.A.

A. E. Clarke, Plant Cell Biology Research Centre, School of Botany, University of Melbourne, Parkville, Victoria 3052, Australia

E. Hacskaylo, Mycology Laboratory, Plant Protection Institute, United States Department of Agriculture, Agricultural Research Service Beltsville Agricultural Research Center, Beltsville, Maryland, U.S.A.

J. Kuijt, Department of Biological Sciences, University of Lethbridge, Lethbridge, Alberta T1K 3M4, Canada

M. E. McCully, Department of Biology, Carleton University, Ottawa, Ontario K1S 5B6, Canada

R. Moore, Department of Biology, Baylor University, Waco, Texas, U.S.A.

J. L. Riopel, Department of Biology, University of Virginia, Charlottesville, Virginia, U.S.A.

M. G. Smart, Department of Plant Pathology, Cornell University, Ithaca, New York, U.S.A.

D. B. Walker, Department of Biology, University of California, Los Angeles, California, U.S.A.

ACKNOWLEDGEMENTS

This book represents the proceedings of a symposium entitled "Vegetative Compatibility Responses In Plants" held at Penn State University in August, 1982. This symposium was supported by the Developmental and Structural Section of the Botanical Society of America and the Developmental Biology Panel of the National Science Foundation.

PREFACE

Plants, like animals, have adapted successfully for survival in environments harboring other organisms that often represent pathological hazards. Animals have responded to these threats by evolving several defense mechanisms, the most notable of which is the immune system. Plants have responded differently, but effectively, in developing compatibility systems that enable them to interact with other organisms. The most intensively studied of these plant systems are the interactions that occur between pollen and stigma. These studies have focussed on mechanisms of cellular recognition, with the finding that glycoprotein surface receptors are involved in the initial cellular interactions that determine compatibility and incompatibility.

The most intensively studied systems involving vegetative compatibility are the interactions that occur between plant pathogens and their hosts. These studies have been especially important in demonstrating the role of lectins in cellular recognition as well as the cellular effects of phytoalexins involved in the hypersensitive response.

Less well studied systems involving vegetative compatibility responses in plants include the interactions between parasitic vascular plants and their hosts, mycorrhizal associations, and graft development. Although superficially unrelated, each of these processes has one feature in common—that two different cells come into direct physical contact, with the repeatable result that they either "accept" (i.e., are compatible with, susceptible to, tolerant of) or "reject" (i.e., are incompatible with, resistant to, intolerant of) each other.

The purpose of this book is to summarize our knowledge of the different compatibility systems characteristic of higher plants. Contributors to this volume have striven to provide a complete review of their respective topics. I hope that this book will not only be informative, but will also stimulate thinking as well as research on vegetative compatibility responses in plants.

TABLE OF CONTENTS

Vegetative Compatibility Responses in Plants

I

Tissue Compatibility and the Haustoria of Parasitic Angiosperms

Job Kuijt

Department of Biological Sciences, University of Lethbridge,
Lethbridge, Alberta T1K 3M4, Canada.

INTRODUCTION

Students of parasitic angiosperms need to accept one vital fact early in their careers: no matter what general subject is under discussion, they usually represent the lunatic fringe. This is so in the evolution of angiosperms in general, as parasites show a very deviant mode of life and are, almost by definition, highly derived organisms. There is also a place for parasitic angiosperms in plant pathology, but it is a rather isolated one. In plant morphology, what stranger roots exist than haustoria? And what embryo sacs match those of Loranthaceae in sheer perverse behavior?

It does not take much thought to appreciate that parasitic angiosperms do indeed have a legitimate role to play in discussions dealing with tissue compatibility. After all, *a normal haustorium is an extraordinarily successful vegetative graft.* In fact, it has been suggested that parasitism had its origin in the type of root grafting which is often seen among the roots of forest trees. Be that as it may, in the course of evolution parasitic plants have "learned" to establish living contact with taxonomically unrelated hosts in such a way as to cause a minimal disturbance in the latter except for the tapping of vascular tissues. Also, there is evidence that some parasites are accomplished manipulators in being able to give direction to developmental processes in adjacent host tissues in such a way as to further favor the parasite. For example, Dr. Inge

1

Dörr at Kiel has unpublished evidence that the endophyte of dodder, *Cuscuta,* is able to induce sieve tube differentiation in adjacent host tissues. I think it is appropriate (and humbling) to remind ourselves that many difficulties in the establishment of heterografts which face us in experimental work today were, in effect, overcome millions of years ago by evolving parasitic plants. That the cells of the endophytic system of dwarf mistletoes (*Arceuthobium*) can live harmoniously in direct contact with those of conifers, for example, is an outstanding evolutionary accomplishment. We thus have, in each parasitic angiosperm, at least two taxonomically diverse individuals living in intimate and harmonious structural and physiological contact.

It is necessary to clarify what we mean by a parasite in the higher plants. We are referring to a plant which, by means of highly specialized organs called *haustoria,* penetrate living host tissues and absorb certain materials from them. There may be one haustorium per parasite (the primary haustorium) or there may be numerous ones, depending on the species of parasite. Thus we find that many mistletoes have a single, frequently massive or extensive haustorium or haustorial system, while most parasitic Scrophulariaceae such as *Castilleja* have numerous very small ones that are more simple in construction. The definition of parasitism in the Angiosperms, therefore, hinges on the concept of the haustorium. Many other plants are often called parasites but, no matter how unusual their mode of life may be—I am thinking of saprophytes, epiphytes, stranglers, and carnivores—, they do not fit into our present discussion. Parasitism in this restricted sense has evolved in some 8 or 9 places in the Angiosperm system (Kuijt 1969).

It is essential at this point that I draw the reader's attention to two salient facts. First of all, what we know of the haustoria of these plants shows an extraordinary and often baffling diversity (Kuijt 1969, 1976, 1977). The haustorium of the dodder (*Cuscuta*), for example, has little in common with that of most other parasites. Even within a group much variation may exist. For instance, the haustorium of a mistletoe such as *Arceuthobium* has nothing in common with that of many other mistletoes such as *Psittacanthus.* My second point is this: it is astonishing to find how little these fascinating organs have been studied in the past, even with the simplest descriptive techniques. There are several entire families where the haustoria are very poorly known indeed. Fewer than half a dozen haustoria have been studied ultrastructurally and even there, with the happy exception of *Cuscuta,* there are major gaps in each case. This combination of ignorance on our part and diversity on the part of the parasites is treacherous when it comes to making generalizations. It is no serious exaggeration to say, when looking at parasitic Angiosperms as a whole, that we know nothing about the physiology of mature haustoria and only a little about the details of their structure.

2

DEVELOPMENT OF
THE HAUSTORIAL CONNECTION

In the establishment of haustorial contact there are, in a rough sort of way, three successive phases, and in each of these phases the problems may be significantly different.

1. There is, first of all, the phase which leads to the earliest physical contact. In the case of a germinating seedling of a parasite, this implies certain recognition phenomena leading to the onset of germination and associated growth, possibly including tropic curvature towards the host organ, and the differentiation of the young haustorial organ in preparation for entrance. Much of this process will be repeated where further haustoria are produced, although the details may be different. I will not refer to this phase—which others may well wish to further subdivide—for mainly two reasons. The first one is that this is the subject of Dr. Riopel's presentation; the second one, that this is where a vast amount of structural diversity exists which, while it may intrigue the specialist, has little if anything to do with the topic of vegetative compatibility responses in plants.

2. The second phase is what we might call the "break and entry" phase. The haustorium is in position against the host organ, poised for action. The parasite now faces perhaps the most difficult problem of all: it must play the aggressor without causing a strong incompatibility response in the host tissues which are invaded. This is a very delicate operation, as some host cells must undoubtedly be killed as the intrusive organ advances. Somehow the parasite must, with a minimum of disturbance to the host, grow through living tissues in which (in most cases) it has no direct nutritional interest, to reach the vascular tissues of the host and establish a conduit through which it can take in materials.

3. The third phase is characterized by a fully established operational haustorium through which materials are diverted to the more remote portions of the parasite. It is a stable phase in the sense that in many parasites no further host tissues are invaded. However, a coordinated growth pattern between host and parasite may still lead to rather dramatic additional developments in those haustoria which are perennial. In some highly advanced groups like Viscaceae and Rafflesiaceae the endophyte does, in fact, invade new host tissues on an apparently continuing basis. In such plants we see what amounts to a prolonged second phase; or perhaps it is more meaningful to say that a distinction between the "break and entry" phase and the mature phase no longer exists.

One of the great difficulties facing us when discussing the fully established haustorium as a whole, as I have already stated, is the fact that (especially functionally) it is very poorly understood. In the haustorium of Santalaceae, for example, we find numerous structural details which seem to make no physiological sense and which, in some cases, even seem to *militate against* an efficient uptake of materials from the host (Kuijt 1977). The anatomical details of parts of the haustorium away from the host-parasite interface are of little direct consequence to vegetative compatibility, even though they can scarcely be ignored in physiological considerations. With regard to the endophyte, however, I think it is necessary that I provide a brief sketch. The endophyte is that portion of the haustorium which is embedded in host tissues; sometimes it is much fragmented and the term endophytic system is more useful.

There are basically two extremes in the endophyte. In its simplest form it is a peg-like or conical structure which breaches the epidermis and cortex of the host to reach the vascular tissues. Such endophytes are common in parasitic Scrophulariaceae (see, for example, Dobbins & Kuijt 1973, Figs. 2-4), in the smaller haustoria of Orobanchaceae (Attawi & Weber 1980), and in several other groups. In some cases an endophyte can become quite massive. In *Boschniakia* (Orobanchaceae), for example, the primary haustorium becomes a rather large, more or less conical structure, the growth of which is synchronized with that of adjacent host tissues. The extreme of this form of endophyte is seen in the enormous haustoria of *Psittacanthus,* which may be nearly a foot in diameter. Here the host-parasite interface forms a system of radially aligned and branching grooves and ridges (Kuijt 1964). The actual anatomy of the haustorial interface of this very common neotropical mistletoe has not yet been studied, but its intricately fluted contours demonstrate a very harmonious coexistence of host and parasite lasting over many years.

In some of the more advanced parasites we find that the endophyte, almost immediately upon entering the host tissues, divides into separate cords or strands which grow parallel to the axis of the host (i.e., within its cortical or phloem tissues). In most cases the more interior side of such *cortical strands* bears radial processes, the *sinkers,* which are anchored in the secondary xylem of the host. Such an endophytic system, with variations, characterizes the mistletoe family Viscaceae, which includes the American *Phoradendron* (Calvin 1968). An extreme form of this type of endophyte is present in dwarf mistletoes (*Arceuthobium*) where the cortical strands of an individual may run for many meters in host tissues, and are fragmented very finely, the ultimate tips of filaments being unicellular and, in some situations, being permanently lodged in the apical meristem of the host (Kuijt 1960). In a quite unrelated family, Rafflesiaceae, a similar endophytic system has evolved but, in contrast to Viscaceae, even vascular tissues seem to have disappeared completely, and a dimorphism between sinkers and cortical strands does not seem to be present (Dell et al. 1982).

DEVELOPMENTAL COORDINATION BETWEEN THE HOST AND PARASITE

What I should like to do now is to focus on the clearest examples of developmental coordination between host and parasite, as this surely is evidence of a very high degree of compatibility. However, it is first necessary to say that such coordination is often not evident. There are numerous annual parasitic plants and perennials with ephemeral haustoria, such as in Scrophulariaceae and some Orobanchaceae. As mentioned earlier, the endophyte of this group tends to be very simple in anatomical structure and, while a remarkable harmony may exist between the two systems, I know of no evidence of developmental coordination. It is in the perennial or long-lived haustoria that we see the clearest evidence of growth coordination. In such organs, growth and differentiation is resumed each growing season, and there is a clear need for mutual adjustment of these activities in the two systems.

The most striking phenomenon here is that a permanent meristem of some type in many mature perennial haustoria may be continuous with the vascular cambium of the host. This cambial continuity is established at a very early time (cf. Sallé 1978, Kuijt 1960). It seems to be a matter of a part of the endophyte penetrating to the host's cambium, where it assumes the latter's characteristics. Cambial continuity is of two very different forms, depending upon the overall structure of the haustorium.

In the more diffuse endophytic systems, the great bulk of the endophyte lies clearly outside the host's vascular cambium. Examples are *Arceuthobium* (Kuijt 1960; Srivastava & Esau 1961), *Phoradendron* (Calvin 1967), and *Viscum* (Sallé 1979). In some Rafflesiaceae, this may be true for nearly the entire endophyte (Dell et al. 1982). From the cortical strands (called *cortical haustoria* by Calvin 1967) numerous radially oriented processes called *sinkers* extend deeper into the host. How sinkers (referred to as "suçoirs secondaires" by Sallé 1979) achieve their anchorage within the xylem is an interesting question, but does not concern us here, and the same is true for the xylem continuity with the host which develops in most (but not all) instances (cf. Srivastava & Esau 1961). What is of importance here is that each mature sinker lies radially across the host's active vascular cambium, and thus stands in danger of being ruptured unless the parasite accommodates for the host's secondary growth. There is evidence in *Phoradendron* that sinkers occasionally become completely separated from the rest of the endophyte (Calvin 1967), possibly because of intrusive growth of fusiform initials of the host along the margin of the sinker, constricting the latter's "neck." A much

5

more common situation, however, is the formation of an intercalary cambial zone across the "neck" of the sinker in the area of the host's vascular cambium (Fig. 1). New tissues are thus added to the sinker, which can now remain in place and maintain its continuity with cortical strands (it should be noted that xylem continuity requires further explanation). At least in *Phoradendron serotinum* there is clear evidence that an annual increment is produced within some sinkers (Calvin 1967). The vascular cambium of the host is therefore continuous with the intercalary meristem of the parasite's sinker. A similar solution may well have evolved in many other parasites with perennial endophytes forming sinkers, as in *Arceuthobium* (Kuijt 1960, Srivastava & Esau 1961) and *Viscum* (Sallé 1979). In the latter case a detailed investigation has shown that the cycle of activity of this meristem approximates that of the host's vascular cambium. In a separate paper the same author (Sallé 1978) has shown the early origin of the intercalary meristem of the sinkers, and the fact that the cycle of sinker initiation also corresponds to the activity cycle of the combined host-parasite meristem system. The sinker systems of other species of Viscaceae often show similar features (Thoday 1956, 1957, 1958b), and intercalary meristems may be expected. However, these haustoria have not been studied in sufficient detail, and show complicating variations as to host species and age of the parasite. The same is true for what appear to be sinkers in several Loranthaceae.

The relative position of the more massive haustorial organs which form graft-like unions in many species can scarcely be maintained by an intercalary meristem. In such systems the parasite instead evolves a continuity between its own vascular cambium and that of its host (Fig. 2). The most striking instances are undoubtedly found in those Loranthaceae and Viscaceae with massive, unified endophytes. Such attachments are often referred to as graft-like, a term indicative of cambial and histological continuity across the host-parasite interface. This phenomenon is a recurrent theme in the long series of papers by Thoday, who in one case also speaks of and illustrates the congruence of the annual rings of the two partners (Thoday 1956). In the *Psittacanthus* woodrose something similar may happen (Kuijt 1970). We may assume that in all these cases the activity rhythms are synchronous on both sides of the interface, reflecting a very high degree of compatibility. A nearly identical host-parasite cambial continuity is present in *Boschniakia,* a perennial genus of Orobanchaceae (Toth & Kuijt, unpublished).

The most highly evolved coordination which exists between parasitic angiosperms and their hosts has nothing to do with lateral meristems, and is found in the synchrony of growth rhythms of certain *Arceuthobium* species and their most common hosts, and in a single species of *Pilostyles* (Rafflesiaceae) (Kuijt 1960). Here the most distal portions of some endophytic filaments are permanently embedded in the apical meri-

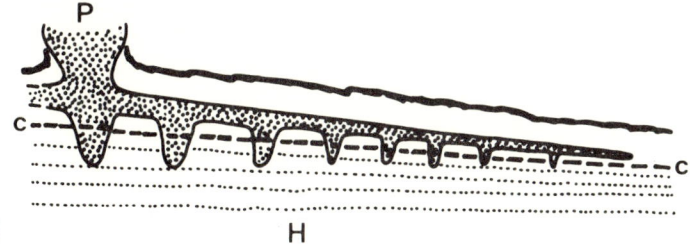

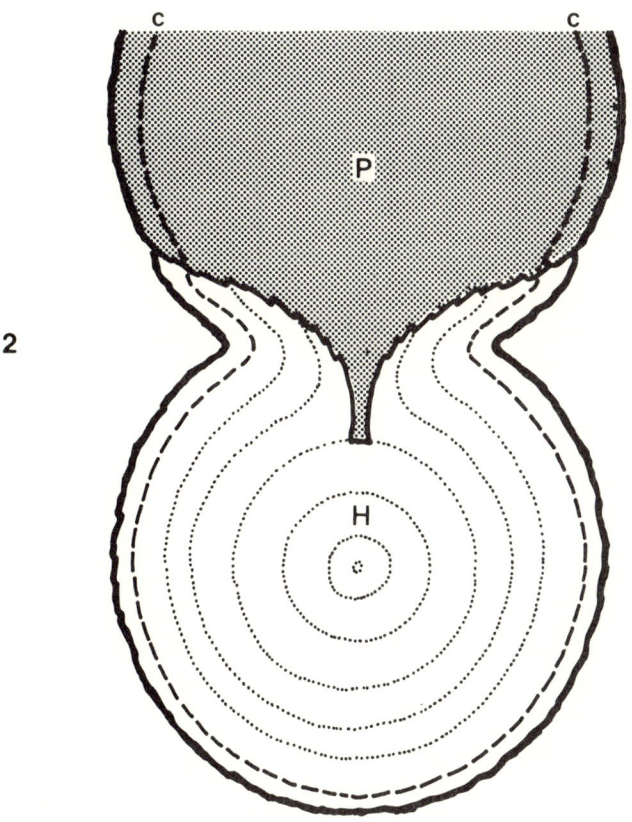

FIGURES 1 & 2. *Diagrammatic representations of the two major types of larger haustoria, showing continuity between the cambia (c) of host (H) and parasite (P). Fig. 1.—The diffuse endophytic system as found in many Viscaceae, with longitudinal cortical strands from which radial sinkers extend downward into the host wood. Fig. 2.—The more massive, unified haustorium as exemplified by many Loranthaceae and some others.*

stem of the host, a situation which can persist only if elongation and apical activity of the host are perfectly synchronized with those of the parasite. I called such situations "isophasic parasitism" many years ago (Kuijt 1960). They are associated with an extremely regular and predictable shoot-emergence pattern from the host, and with certain broom-like formations of the affected parts. The term "systemic broom" has also been applied to such brooms (Tinnin, Hawksworth, & Knutsen 1982) but has somewhat less precision. From our present point of view it is interesting that a species of *Arceuthobium* may demonstrate this remarkable degree of developmental harmony with one host species but not with another (Kuijt 1960).

The haustorium of the dodder, *Cuscuta*, deserves a few separate comments. This unique organ has been elucidated in several papers by Dr. Inge Dörr of Kiel (Dörr 1967, 1968a, 1968b, 1969, 1972; Israel et al. 1980) and I am indebted to Dr. Dörr for several valuable suggestions. The endophyte of *Cuscuta* has a multicellular axis embedded in the outer parenchyma (cortical) tissues of the host stem or leaf. From the tip of this axis, slender uniseriate filaments extend in various directions, often growing through host cells which remain alive. In principle these filaments (often called search hyphae or Such-hyphen) are of three types. First of all, there are hyphae which do not develop contact with the host's vascular tissues and normally remain strictly parenchymatous. A second type reaches the host xylem and establishes a xylem bridge through differentiation of xylary elements. A third type differentiates cells similar to but different from sieve elements when establishing contact with the host's phloem.

HOST RECOGNITION AND SPECIFICITY

There can be little doubt that host recognition at whatever level is a subject relevant to compatibility responses in plants. Riopel and his co-workers have shown that some parasites recognize host roots. The questions I would like to raise are these: is there any evidence that the endophytic "hyphae" of *Cuscuta* can discriminate between tissue types of the host, or are the three hyphal types merely a consequence of the tissues which happen to be reached?

The answers which are available to these questions are by no means clear-cut. First of all, the parenchymatous hyphae probably retain the potential to differentiate into either of the other two. Confusing the situation is the fact, as Dr. Dörr tells me, that a differentiated xylem hypha sometimes has no contact with host xylem, ending blindly inside a parenchyma cell. However, it appears that hyphae which do reach a vascular tissue differentiate in accordance with it. This, of course, still is no evidence for tissue recognition. In fact, there seems to be no evidence of

8

young, developing hyphae sensing the position of either of the two vascular tissues and "purposefully" growing towards them. In this sense there may not be recognition at a distance, and the term "search hyphae" may, indeed, be somewhat misleading.

If we study the direction of differentiation in xylary and phloic hyphae we discover a striking contrast. Xylary differentiation begins at the host xylem, working its way back to achieve continuity (Thomson 1925; cf. Kuijt 1969, Fig. 7-23). That is, it is basipetal. We may thus assume that the morphogenetic signal emanates from the host xylem. In phloic hyphae, in contrast, differentiation is acropetal. Differentiation (presumably when the hyphal tip is already in the vicinity of the host's sieve tube members) starts in the *Cuscuta* shoot and follows the hypha until the host phloem is reached. Surely the only way that this process can take place is through the reception, by the proximal cells of a hypha or tissues even more deeply situated in the *Cuscuta* system, of a specific signal which originates in the host phloem. The most obvious suggestion as to the nature of this signal is the concentration of sucrose which young hyphae might already be absorbing and transporting before their cells differentiate. Be this as it may, there appears to be little doubt that such a hypha is able to sense the type of vascular tissue contacted by its tip. In other words, we seem to have a clear case here of tissue recognition at a distance.

Perhaps the ultimate visual evidence of compatability between cells of different genotypes would be a continuity of protoplasmic membranes via plasmodesmata. In older literature, several unsupported (or inadequately supported) statements to this effect do indeed appear but, since this work predates electron microscopy, they cannot be relied upon (Kuijt & Toth 1976). In recent years, however, three separate claims of this type have been made by workers using the techniques of electron microscopy, and these claims must be carefully evaluated. Dörr (1968b, 1969) has shown rather convincingly that comparable connections sometimes exist between dodder, *Cuscuta,* and its host. However, they occur only under special circumstances, and are at best evanescent. In a dwarf mistletoe, *Arceuthobium pusillum* Peck, similar connecting strands are claimed to exist (Tainter 1971), but I believe in this particular case that the possibility of an error in cell identification cannot be ruled out.

More recently, a study of the endophyte of the Australian *Pilostyles hamiltonii* C. A. Gardner (Rafflesiaceae) has appeared (Dell et al. 1982) in which it is claimed that plasmodesmata are abundant between cells of parasite and host, *Daviesia* (Leguminosae). An electron micrograph (Fig. 10) purports to show a cluster of plasmodesmata connecting the living cells of the two partners. Although the photograph is suggestive and plasmodesmata are undoubtedly correctly identified, the latter are cut transversely and are not seen to connect the protoplasm of the two cells.

9

I submit, therefore, that the evidence of a protoplasmic bridge between host and parasite even in this case lacks substance. I should add that I am currently engaged in a similar study of the Californian *Pilostyles thurberi* Gray on *Dalea emoryi* Gray (Leguminosae) which shows no evidence whatever of trans-interface plasmodesmata (Kuijt & Bray, unpublished).

In summary, we must be extraordinarily cautious in this question, both with regard to identification of host and parasite cells and to the documentation of full-length plasmodesmata. As I have mentioned elsewhere (Kuijt & Toth 1976), current views scarcely facilitate the differentiation of plasmodesmata on any but division walls. It is illuminating that a parallel controversy exists with regard to the interface between grafted tissues. There is no truly reliable evidence anywhere in the literature of proto-plasmic connections across a graft union, or in graft hybrids such as *Laburnocytisus* (Leguminosae).

A final note seems justified on the question of host specificity which, at a very different level, is also a reflection of vegetative compatibility. I dealt with this topic at some length in an earlier paper (Kuijt 1979). Certainly, in numerous other parasitic organisms specificity is highly developed. The rust fungi (Uredinales), for example, are notoriously specific in their choice of hosts. For animals, Caullery (1952) writes: "One of the characteristics of parasitism . . . is the specificity of these associations; they always occur between definite species." More recently we read the following statement by the plant pathologist Sequeira (1978): "Obligate plant pathogens . . . exhibit a great deal of specificity and can grow only in certain varieties of hosts."

The corollary assumption that a high degree of host specificity exists in parasitic angiosperms, however, turns out to be false in most cases. Parasitic angiosperms tend to be very poor taxonomists, indeed. The Eurasian mistletoe *Viscum album* L. in Europe alone has been recorded from more than a hundred genera of hosts, for example. Interestingly enough, its close relative *V. cruciatum* Sieb. ex Spreng. is known only from a single host. This illustrates the fact that, where such specificity occurs, it is a sporadic phenomenon characterizing individual, small taxa. It must be admitted, however, that intriguing local variations in host preference seem to occur in some mistletoes which may indicate that the situation is more complex than here indicated. The basic facts are rarely totally reliable because of the weaknesses inherent in the source of information, whether based on observations in the field or in the herbarium. Furthermore, there is little or no evidence that host specificity increases with evolutionary advancement, as often seems to be the case in other groups of parasites (for a somewhat contrasting point of view, see Barlow 1966).

The question of the relationship between tissue compatability and host preferences, therefore, has limited meaning vis-á-vis the information available. It is true that certain parasites have never been found on certain host species even where abundant opportunity exists. No *Arceuthobium*, for example, has ever been collected from an angiosperm host. We may presume, considering the coexistence of angiosperm shrubs or trees with *Arceuthobium* spp. in many places, that the basis for this phenomenon ultimately lies in tissue incompatability. However, this is not a foregone conclusion, and needs to be verified experimentally. Needless to say, nothing of the sort has been done with any parasitic angiosperm. This cannot be blamed only on botanists, as most parasitic angiosperms are remarkably uncooperative from the point of view of experimental work.

LITERATURE CITED

Attawi, F. A. J., and H. C. Weber. 1980. Zum Parasitismus und zur morphologisch-anatomischen Strucktur der Sekundärhaustorien von *Orobanche*-Arten (Orobanchaceae). Flora *169*:55-83.

Barlow, B. A. 1966. A revision of the Loranthaceae of Australia and New Zealand. Austral. Jour. Bot. *14*:421-499.

Calvin, C. L. 1968. Anatomy of the endophytic system of the mistletoe, *Phoradendron flavescens*. Bot. Gaz. *128*:117-137.

Caullery, M. 1952. Parasitism and symbiosis. Sidgwick & Jackson Ltd., London.

Dobbins, D. R., and Job Kuijt. 1973. Studies on the haustorium of *Castilleja* (Scrophulariaceae). II. The endophyte. Can.J. Bot. *51*:923-931.

Dörr, I. 1967. Zum Feinbau der "Hyphen" von *Cuscuta odorata* und ihrem Anschluss an die Siebröhren ihrer Wirtspflanzen. Naturwissenschaften *54*:474.

————. 1968a. Zur Lokalisierung von Zellkontakten zwischen *Cuscuta odorata* under verschiedenen höheren Wirtspflanzen. Protoplasma *65*:435-448.

————. 1968b. Plasmatische Verbindungen zwischen artfremden Zellen. Naturwissenschaften *55*:396.

————. 1969. Feinstruktur intrazellular wachsender *Cuscuta*-Hyphen. Protoplasma *67*:123-137.

————. 1972. Der Anschluss der *Cuscuta*-Hyphen an die Siebröhren ihrer Wirtspflanzen. Protoplasma *75*:167-184.

Israel, S., I. Dörr, and R. Kollmann. 1980. Das Phloem der Haustoria von *Cuscuta*. Protoplasma *103*:309-321.

Kuijt, Job. 1964. Critical observations on the parasitism of New World mistletoes. Can. J. Bot. *42*:1243-1278.

————. 1969. The biology of parasitic flowering plants. University of California, Berkeley & Los Angeles.

————. 1970. Seedling establishment in *Psittacanthus* (Loranthaceae). Can. J. Bot. *48*:705-711.

————. 1979. Host selection by parasitic angiosperms. Symb. Bot. Upsal. *22*:194-199.

————, and R. Toth. 1976. Ultrastructure of angiosperm haustoria—a review. Ann. Bot. *40*:1121-1130.

11

Sallé, G. 1978. Origin and early growth of the sinkers of *Viscum album* L. Proto-plasma *96*:267-273.

—————. 1979. Le système endophytique du *Viscum album*: anatomie et func-tionnement des suçoirs secondaires. Can. J. Bot. *57*:435-449.

Sequiera, L. 1978. Lectins and specificity. Ann. Rev. Phytopath. *16*:453-481.

Srivastava, L. M., and K. Esau. 1961. Relation of dwarf mistletoe (*Arceuthobium*) to the xylem tissue of conifers. I. Anatomy of parasite sinkers and their con-nection with host xylem. Amer. J. Bot. *48*:159-167.

Tainter, F. H. 1971. The ultrastructure of *Arceuthobium pusillum*. Can. J. Bot. *49*:1615-1622.

Thoday, D. 1951. The haustorial system of *Viscum album*. J. Exp. Bot. *2*:1-19.

—————. 1956. Modes of union and interaction between parasite and host in the Loranthaceae. I. Viscoideae, not including Phoradendreae. Proc. Roy. Soc. B. *145*:531-548.

—————. 1957. Ibid., II. Phoradendreae. Idem *146*:320-338.

—————. 1958a. Ibid., III. Further observations of *Viscum* and *Korthalsella*. Idem *148*:188-206.

—————. 1958b. Ibid., IV. *Viscum obscurum* on *Euphorbia polygona*. Idem *149*:42--57.

—————. 1960. Ibid., V. Some African Loranthoideae. Idem *152*:143-162.

—————. 1961. Ibid., VI. A general survey of the Loranthoideae. Idem *155*:1-25.

—————. 1963. Ibid., VII. Some Australian Loranthoideae with exceptional fea-tures. Idem *157*:507-516.

Tinnin, R. O., F. G. Hawksworth, and D. M. Knutson. 1982. Witches broom formation in conifers infected by *Arceuthobium* spp.—an example of para-sitic impact upon community dynamics. Amer. Midl. Nat. *107*:351-359.

II

The Biology of Parasitic Flowering Plants: Physiological Aspects

J. L. Riopel

Department of Biology, University of Virginia
Charlottesville, VA

INTRODUCTION

It is the intent of this paper to consider some of the physiological aspects of early host-parasite relationships in angiosperm root parasites. Some of these plants are of great economic significance, and it is not surprising that much has been written about them. Most of the earlier work is cited in Subramanian and Srinivasan, 1960; King, 1966; Beilin, 1968; and Kuijt, 1969. More recent reviews are those by Kuijt 1977, 1979; and Musselman, 1980, 1982. Root parasite plants may be faculative or obligate. Many photosynthetic species can be grown in pot cultures without a host (e.g., *Agalinis aphylla, Aureolaria flava, Dasistoma macrophylla, Macranthera flammea, Seymeria cassioides* and others; see Mann and Musselman, 1981), but it is likely that under normal field conditions, host attachment is the usual mode. Indeed, it is rare if ever that field collections reveal plants that are devoid of attachments. Thus, the parasitic mode implies that the early developmental events of seed germination through the establishment of functional host connections are uniquely important, and we can anticipate that these plants have evolved in significant ways to find a suitable host plant quickly and efficiently. There are numerous studies of these early stages, but much is still to be learned. For the purpose of this review, these stages are divided into germination, regulation of haustoria, and haustorial attachment.

GERMINATION

General Aspects

It has been emphasized by Kuijt (1969) that the germination require-ments of many root parasites are unspecialized. Seeds of many of the hemiparasites germinate when moistened, or if they are a temporate species, require only vernalization. For these plants, germination does not require the host.

For other species germination is far more complex, and it is evident that this stage, as well as other early developmental events, are cued to specific substances that signal the presence of the host root system. The requirement for host presence for germination has been reported for *Orobanche* (Vaucher, 1823; Koch, 1887; Chabrolin, 1938; and others), *Tozzia* (Heinricher, 1901), *Aegenetia*, (Kusano, 1908), *Striga* (Saunders, 1933), and *Alectra* (Botha, 1948, 1950). As shown in Table 1, *Striga* and *Orabanche* are serious parasites of several food crops, and most of our information on germination physiology concerns these species. Earlier reviews on germination are those by Kadry and Twefic (1956), Sunder-land (1960), Brown (1965), and Edwards (1972).

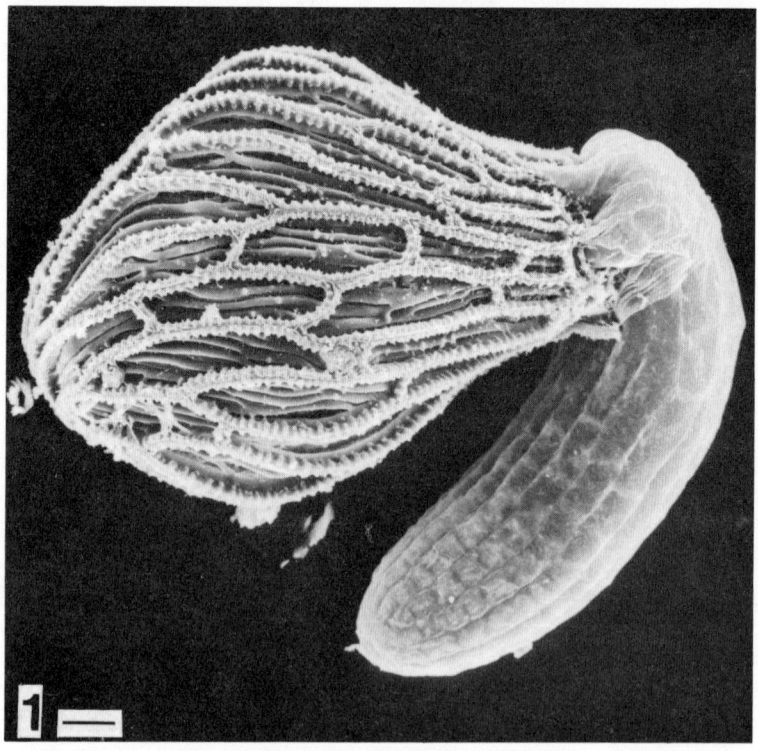

FIGURE 1.   Striga asiatica *12 hrs after germination. (Scale 20μm)*

14

TABLE 1. *Striga* and *Orobanche* species of economic importance (Musselman, 1980)

Name and synonym	Geographic area	Major crop host(s)
Striga hermonthica (Del.) Benth. *S. senegalensis* Benth.	Africa	Millet, sorghum, rice, maize
S. asiatica (L.) Kuntze *S. lutea* Lour. *S. hirsuta* Benth. *S. coccinea* Benth.	Africa; Arabia through the Indian subcontinent to China, Indonesia, and the Philippines; USA (Carolinas)	Millet, sorghum rice, maize, sugarcane
S. gesnerioides (Willd.) *Vatke* *S. orobanchoides* Benth.	Africa, India; USA (Florida)	Cowpea, tobacco, sweet potato
S. curviflora Benth.	Australia	Sugarcane
S. parviflora Benth.	Australia	Sugarcane
S. latericea Vatke	Ethiopia	Sugarcane
S. densiflora Benth.	India	Sorghum, sugarcane
S. forbesii Benth.	East Africa	Maize
S. aspera (Willd.) Benth.	Africa	Sorghum
S. euphrasioides Benth. *S. angustifolia* (Don) Saldhana	India	Sorghum, sugarcane
S. barteri Engl.	West Africa	Rice
S. brachycalyx Skan.	West Africa	Sorghum
Orobanche ramosa L. *O. mutelii* F. W. Schultz *Phelypaea ramosa* L. C. A. Mey	Southern and Central Europe; introduced elsewhere, e.g. France, Cuba, Mexico, USA (California)	Tomato, tobacco, potato, hemp
O. cernua Loefl. *O. cumana* Wallr. *O. brassicae* Novopokr.	Southern Europe and Western Asia	Sunflower, tobacco
O. crenata Forsk. *O. speciosa* DC.	Southern Europe; Mediterranean region	Broadbeans, peas, lentils
O. minor Smith *O. picridis* F. W. Schultz	Europe; introduced in New Zealand, US, East Africa and perhaps elsewhere	Clover, tobacco
O. aegyptiaca Pers. *Phelypea aegyptiaca* (Pers.) Walp.	Central southwestern Asia, Middle East	Broadbeans, melons tobacco
O. ludoviciana Nutt.	United States	Tomato, tobacco

The seeds of *Striga* and *Orobanche* are very similar in appearance (Fig. 1). They are brown, small, usually about .25-3 mm long, and produced in abundance. A single *O. crenata* plant produces as many as

15

40,000 seeds (Kadry and Twefic, 1956). Following a period of after-ripening, seeds of *Striga asiatica* have been reported viable for up to 20 years (Saunders, 1933) and *O. crenata* seeds were still viable after 10 years (Kadry and Twefic, 1956).

Inside the testa, there is usually a single aleurone layer surrounding a thin endosperm. This may be resorbed by the embryo during seed maturation. The embryo is globular. In *O. minor,* Sunderland (1960) described differences in the size and staining density of the cells, but there is no organ differentiation. In the presence of the host, germination can be very rapid. Radicle emergence for *S. asiatica* occurs in 24 hrs (Fig. 1) and within 96 hrs for *Orobanche* (Brown, 1965). The response is effected principally by cell expansion although some cell division also occurs (Sunderland, 1960; Brown, 1965). Both *Striga* and *Orobanche* are obligate parasites. Their seed biology is ideally suited for this lifestyle. Reproductive effort is spent on the production of thousands of very small seeds. They remain viable for long periods and yet, when moistened for a time, they can respond to a host signal and germinate in 1-3 days. For some species, the duration of seed exposure to host stimulant can be brief. *Orobanche speciosa* seeds germinated after only a 30 sec exposure to host exudate (Chabrolin, 1938). Brown and Edwards (1944) have reported similar exposure observations for *S. asiatica.* Seeds germinate only within a well-defined zone not reported to exceed 2 cm from the host root surface (Chabrolin, 1938; Brown and Edwards, 1944). This zone is still larger than the radicle extension limits of about 2 mm. Thus, many more seeds germinate than can extend to the host root surface (Sunderland, 1960).

Sunderland (1960) has shown that in corn roots the germination stimulant is most abundant in the zone just proximal to the root tip. In agar cultures of corn with *S. asiatica,* we observed progressive germination and the establishment of approximately a 2 cm zone during elongation of the host root. The zone is not significantly extended by the production of stimulant as the root matures and is enlarged only by the intrusion into the zone by new main or lateral host roots.

Regulation

Root parasites that are host dependent for seed germination have several factors in common (Fig. 2).

STAGES IN HOST-DEPENDENT GERMINATION

SEED DISPERSAL	AFTER-RIPENING	WATER CONDITIONING	HOST SIGNAL	RADICLE EMERGENCE
	MONTHS TO YEARS	DAYS	HOURS	24-96 HOURS

FIGURE 2. Stages in host-dependent germination

Dry seeds collected from capsules usually require an after-ripening period of several months. Germination frequency then appears to improve for several years of dry storage at room temperatures (Vallance, 1950). There are some indications that low temperatures storage may also improve germination (Edwards, 1972).

An interesting aspect is that seeds of *Striga* and *Orobanche* must be soaked or conditioned in water for several days before maximum response to host stimulant occurs (Brown and Edwards, 1944). Seeds of *Alectra vogelii* have a similar requirement (Botha, 1950; Okonkwo, 1975).

Brown (1965) has shown that germination frequencies peak for *Striga* around 10 days of conditioning, while seeds of *Orobanche* are stable in the conditioning period for months. With prolonged conditioning, Vallance (1950) reported a second elevation in germination frequency for *Striga hermonthica*. In general, there are very few studies of the effects of longer pretreatments. Vallance found that seeds remain viable when dried after pretreatment, a situation not unlike the effects on seeds during normal fluctuations of the water content in soils.

The physiological basis of water conditioning is not known. The duration of the treatment far exceeds the time requirements for normal seed inbibition and the related requirements for the establishment of seed permeability. What we do know about the process is that: (1) Altered CO_2/O_2 ratios during conditioning and exposure to host exudate suggest the occurrence of significant metabolic changes during the process (Vallance, 1951), (2) Spontaneous germination takes place in all species studied (with the possible exception of *O. crenata* (Hiron, 1973) and varies from a usual 0.5-2% during pretreatment to, in one report, as much as 40% (Vallance, 1950), (3) Exposure to gibberellin during pretreatment of *O. ramosa* (Abu-Shakra, Miah, and Saghir, 1970), and *O. crenata* (Hiron, 1973; Garas, Karssen, and Bruinsma, 1974) increases the germination frequency when host root exudate is applied afterwards, (4) Seeds of *S. asiatica* treated with kinetin germinate without conditioning (Yoshikawa, Worsham, Moreland, and Eplee, 1978). With this treatment, germination occurs in three days. Seedlings will develop, but in the continued presence of kinetin, shoot development proceeds without normal radicle extension (Riopel, unpublished), (5) Germination inhibitors have been reported in some seeds (Kust, 1966, *Striga;* Edwards, 1972, *O. crenata*) but not in others (Botha, 1950, *Alectra vogelli*). However, unequivocal evidence for the involvement of inhibitors in regulation has not been shown. It has been suggested that the pretreatment period may be essential for the accumulation of the host germination stimulant or a related substance to threshold levels (Brown and Edwards, 1944; Botha, 1950; Brown, 1965). Proof for this, along with consideration of the possible interactions with endogenous inhibitors must await identification of the functionally active molecules of host root exudates.

Host Root Stimulant

A host stimulant requirement for certain root parasites has been known for a long time (Heinricher, 1898; Pearson, 1912; Saunders, 1933). An understanding of the process has been the focus of several excellent research efforts including studies by A. R. Saunders, R. Brown, A. D. Worsham, and W. G. H. Edwards. Unfortunately, the isolation and characterization of the active principle has not been achieved. Problems encountered include low concentrations contained in root exudate, the possibility of synergism between several components, the need for absolute purity of sufficient amounts for spectral analysis, variable production by roots, and instability of the active fraction. The chemistry has been reviewed by Brown (1965), Edwards (1972), and Hiron (1973). Some facts about this fascinating puzzle are:

(1) Many compounds have activity in promoting seed germination. A partial list is summarized in Table 2. Clearly, not all of these substances are biologically significant under field conditions. Information on their effects has been helpful in understanding the germination process and has additionally been useful in learning about the influence of compounds like ethylene, now a useful method of controlling *Striga* in the United States (Eplee, 1982).

(2) We may characterize seed germination for some of these plants as host recognition, but only in a general way. Indeed, root exudates from many plants that are nonhosts contain active substances (Heinricher, 1898; Pearson, 1912) and the planting of false hosts or trap crops is a control strategy in some areas (Musselman, 1980).

(3) Substances even remotely useful as molecular recognition cues are not likely to be the usual hormones or metabolites common to many

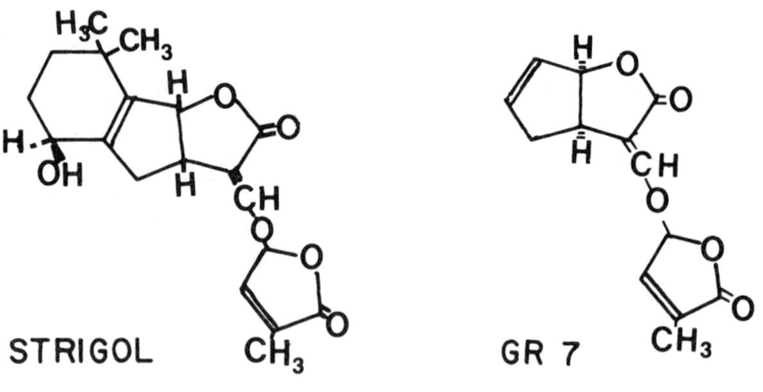

FIGURE 3.   *Strigol and analog GR7*

exudates. The finding that strigol, purified from cotton, a nonhost of *Striga*, and certain analogues (Fig. 3), promotes germination in *Striga*

TABLE 2.   Chemical Treatments Influencing Seed Germination of Parasitic Plants

Compound	Plant	Authors	Effects
indoleacetic acid	O. crenata	Edwards et al. 1976	pretreatment enhances response
	O. crenata	Garras et al. 1973	inactive
	Aeginetia indica	French et al. 1976	stimulated germination
	S. asiatica	Egley 1972	inhibited germination
gibberellic acid	O. hederae	Privat 1960	stimulated germination
	O. ludoviciaena	Nash and Wilhelm 1960	increased over spontaneous germination
	O. ramosa	Abu-Shakra 1970	conditioning enhanced, low promotion in absence of host exudate
	S. asiatica	Egley 1972	stimulated scarified seed
	O. crenata	Garras 1973	aids conditioning
	O. crenata	Edwards et al.	reduces minimum exposure time to host stimulant
	S. asiatica	Cook et al. 1966	no effect
	Aeginetia indica	French and Sherman 1976	promotion
kinetin and related 6-substituted purines	O. crenata	Garras et al. 1973	inhibition
	S. asiatica	Worsham et al. 1959	promotion
		Yoshikawa et al. 1978	promotion in culture
	O. crenata	Edwards et al. 1976	inhibition
	Aeginetia indica	French and Sherman 1976	no effect
ethylene	S. asiatica	Egley and Dale 1970	promotion
	O. crenata	Edwards et al. 1976	no effect
coumarin 4-hydroxy- coumarin scopeleton	O. crenata	Edwards et al. 1976	inhibition
	S. asiatica	Worsham et al. 1962	promotion
thiourea and allythiourea	S. asiatica	Brown and Edwards 1945	promotion at high concentrations
xyloketose	S. asiatica	Brown et al. 1951	occasional promotion; not confirmed
proline ornithine	O. ramosa	Donini 1959	promotion, not confirmed in other studies
strigol and/or analogs	S. asiatica	Cook et al. 1966	promotion
	S. asiatica	Hsiao et al. 1981	retards conditioning, promotes germination
	O. ramosa	Saghir et al. 1980	promotion
	O. minor	Spelce and Musselman 1981	promotion
sodium or calcium hypochlorite	Aeginetia indica	French and Sherman 1976	promotion
	Alectra vogelii	Okonkwo and Nwoke 1975	promotion

19

and *Orobanche* species, may be significant (Cook et al., 1966, 1972; Pavlista et al., 1979; Saghir et al., 1980; Spelce and Musselman, 1981). Strigol is active at 10^{-11}M (Cook et al., 1966). Compounds of this kind may be present in host root exudates with genus or family specificity, thus providing a very useful recognition cue.

Strigol has not been isolated from corn. We are presently looking at the exudates of both corn and sorghum. As yet, we can add very little to the picture. The germination complex in corn is very nonpolar and is unstable even in partially purified form. We have no evidence that Strigol is the corn germination stimulant. The sorghum stimulant may be yet another compound. It is more polar and it is quite stable to at least a 10^4 fold purification from crude exudate. We anticipate therefore, that structural characterizations can be made, but also that there are several root exudate molecules active in the *Striga* and *Orobanche* systems. With *Orobanche* there is also some recent evidence that a benzopyran derivative is stimulatory (Davis et al., 1977 and Davis et al., 1978).

We have yet to learn the full importance of seed germination in host recognition. It seems likely that host signals may be very important in species with very narrow host limits like *Conopholis americana* (squawroot), found only in association with oak trees. For this plant, we have found many germinated seeds and early host penetration stages in the field (Fig. 4) (Baird and Riopel, 1980), but the requirements for germination remain unknown. We have been unable to consistently germinate seeds in the laboratory even in the presence of oak root exudate. The molecular cues that stimulate germination in *Striga* and *Orobanche* are obviously less precise. However, the germination signal is important and clearly operative under field conditions. As suggested by Brown (1965), the later stages of haustorial initiation and haustorial attachment and penetration may also have an important role in the host selection process.

REGULATION OF HAUSTORIA

There are frequently few or no haustoria present in cultures of hemi- or holoparasites grown without a host. Observation of this in cultures of S. *senegalensis* led Okonkwo (1966) to suggest this may be indirect evidence that a stimulus for the formation of haustoria comes from the host root. To test this idea, we have grown a variety of these plants in culture without hosts. Our findings suggest there is a species gradient in the requirement for an exogenous host signal. In culture, *Aureolaria pedicularia* forms abundant haustoria without a host (Fig. 5). Spontaneous haustoria occur in *Agalinis purpurea* in about 2% of the population in young cultures (3-6 wk) and about 20-25% with older ones (2-4 months). We

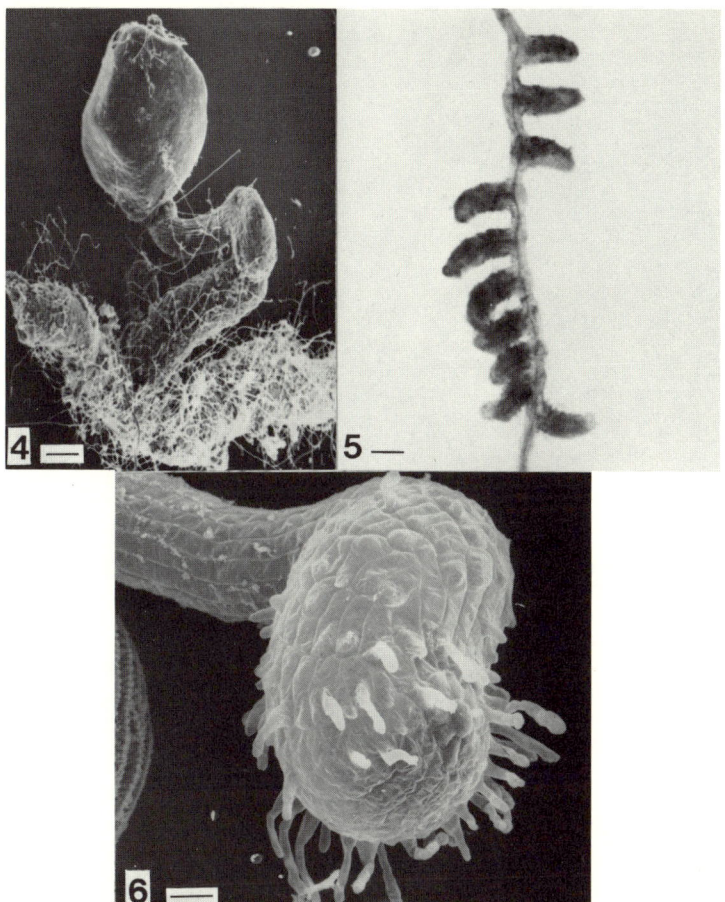

FIGURE 4.   Conopholis americana *seedling attached to* Quercus mycorrhizal
root. Field collected. (Scale 20μm)
FIGURE 5.   *Spontaneous haustoria of* Aureolaria pedicularia *grown in sterile*
culture. (Scale 1 mm)
FIGURE 6.   *Primary haustorium of* Striga asiatica. *24 hrs post-induction. (Scale*
20μm)

have found a lot of variation in the frequency of haustoria in nutritionally
stressed older cultures.

In *Striga*, the requirement for host signal seems absolute. This was
observed for *S. senegalensis* (Okonkwo, 1966) and we have observed
that *S. asiatica* cultures never form haustoria in the absence of a host
stimulus. However, when exposed to host signal, roots devoid of haustoria form them overnight (Fig. 6). The cellular response is as rapid as for
any multicellular organ in angiosperms (Riopel, 1979). In *Agalinis*, root
elongation slows temporarily (Fig. 7) and typically, just proximal to the
root apex, cortical cells enlarge radially followed by cell divisions in the

21

ROOT ELONGATION

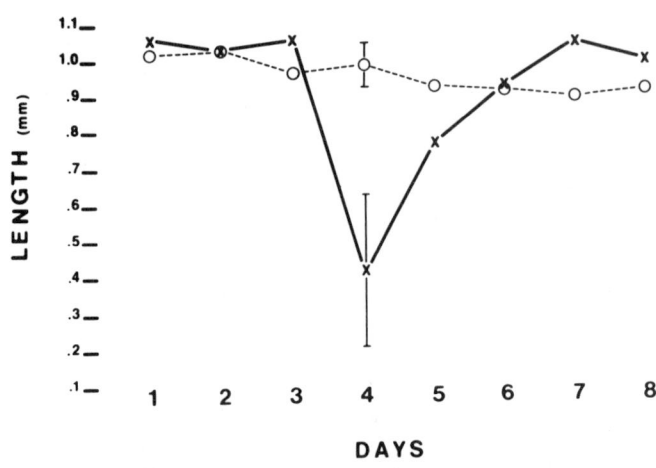

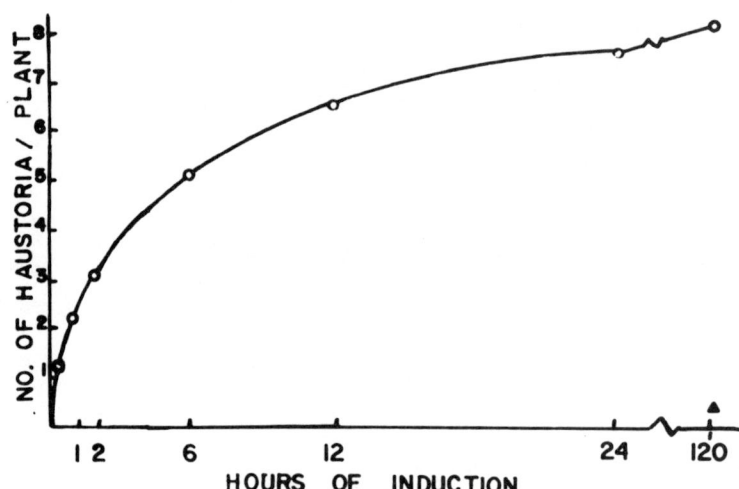

FIGURE 7. *Comparison of normal root elongation (o) and elongation during haustorial initiation (x) in* Agalinis purpurea

FIGURE 8. *Effect on haustorial promotion by different durations of exposure to host exudate. Significance 95% confidence level, F test. ▲=control level 0.5 haustoria/plant*

epidermis and cortex. The development of hairs from epidermal cells follows (Baird and Riopel, 1982).

To induce haustoria, the minimum exposure time of *Agalinis* roots to host exudate is about 30 min. Figure 8 compares the haustorial frequency of plants exposed for different time intervals to host exudate, rinsed and grown in water for 5 days with plants continually exposed to exudate. These results also show a saturation effect of host signal. At 24 hrs, no significant increase in the frequency of haustoria/plant occurs with longer or even continuous exposure.

What is the nature of the stimulant? A variety of substances contained in most exudates are inactive (Riopel, 1979), yet in *Agalinis* and in *Orthocarpus purpurascens* studied by Atsatt et al. (1978), host root exudate, plant extracts, and a number of plant products promote haustorial formation. A few years ago, we observed that gum tragacanth, a foliar exudate of *Astragalus* spp. (Leguminaceae) was found to be especially potent (Riopel, 1979). We have isolated two active components from tragacanth called xenognosin A 1 and B 2 (Lynn et al., 1981; Steffens et al., 1982). They are flavonoids and bear similar hydroxyl and methoxyl substitutions. The relevance of the substitution pattern to inducer activity became evident when formononetin 3 was identified as a component co-purifying with xenognosin B. Formononetin does not produce haustoria in *Agalinis* at any concentration. Its structure differs only in lacking the 2′ hydroxyl of xenognosin B. We have probed the specificity of the xenognosin A with analogs (Steffens et al., 1982; Kamat et al., 1982). With this molecule, there are two structural features required for activity:

(1) A *meta* relationship of hydroxyl and methoxyl groups, and
(2) an alkyl branching ortho to the methoxyl substituent.

1 Xenognosin A

2 Xenognosin B

3 Formononetin

Xenognosin A 1, which induces haustoria in 86% of the *Agalinis* population, satisfies both of these requirements (Fig. 9). Complete methylation (4a) or demethylation (4b) of xenognosin A eliminates activity, and synthetic derivatives lacking the *meta*-methoxyphenol functionality (4c-e) show no activity in either solution or disc assay. Very significant activity is retained when we modify the disubstituted aromatic ring of xenognosin A as in 4f and 4g. In assays conducted on agar, the activity of both analogues is indistinguishable from xenognosin A. However, when roots are tested with these compounds in solution in depression wells, we found that 4f and 4g (at 10^{-5}M) have less inducing activity than xenognosin A (Fig. 9).

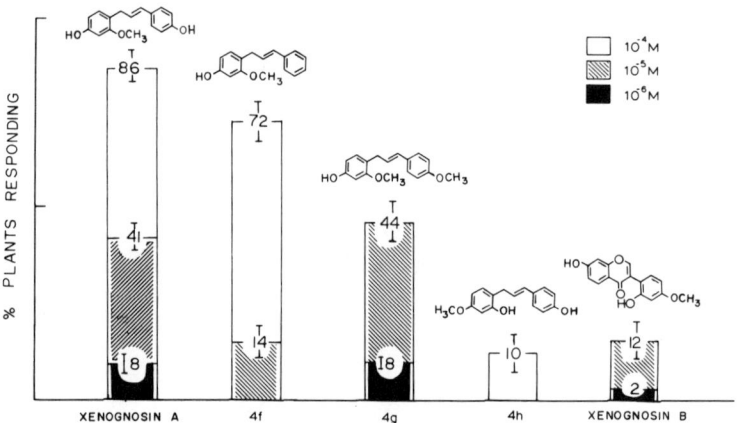

FIGURE 9. *Relative activities of haustorial inducers when presented in solution to 2- to 3-week-old plants of* Agalinis purpurea. *Plants developed an average of two haustoria each when presented with xenognosin A at 10^{-4}M. ▨ , 10^{-4}M; □ , 10^{-5}M; ■ , 10^{-6}M*

The importance of alkyl branching *ortho* to the methoxyl group on the *meta*-methoxyphenol ring is indicated by the activity of xenognosin B 2 and the isomeric xenognosin A (4h). Although xenognosin B possesses considerable activity in an agar disc assay, when presented in solution, its activity is weak (12 percent at 10^{-4}M) and is comparable to the diminished activity of 4h (Fig. 9). Analogs have not been synthesized to test the activity of 2- or 5-substituted 3-methoxyphenol systems, but the importance of the alkyl substituent is substantiated by the observation that *meta*-methoxyphenol 5 is only weakly active in solution.

Further evidence for the importance of the propene side chain oxidation level in xenognosin A comes from activity studies of dihydroxenognosin A 6 which is inactive. We have also tested other synthetic samples prepared by Professor Tsutomu Furuya (Kitasato University, Japan) in his biosynthetic studies of echinatin, 7 (Ayabe and Furuya, 1981). Although many of these compounds contain the requisite regiochemical aromatic substitution of xenognosin A and show activity in disc assay, no

4

R₁ / R₂ / R₃ table:

	R_1	R_2	R_3
a)	OCH_3	OCH_3	OCH_3
b)	OH	OH	OH
c)	H	OCH_3	OCH_3
d)	H	OH	OCH_3
e)	OH	OH	OCH_3
f)	OH	OCH_3	H
g)	OH	OCH_3	OCH_3
h)	OCH_3	OH	OH

5

6

7

compounds with a differently functionalized propene unit show activity in depression slides.

In view of the broad host range of *Agalinis,* the structural specificity of the xenognosins does not imply that they are functionally unique molecules. That they are flavonoids is of interest in view of Atsatt's idea that parasitic plants, like herbivorous insects, may key on a host's defense chemicals as recognition cues (Atsatt, 1977). Xenognosin A and B belong to a family of compounds attributed *in situ* with roles as constitutive antibiotics and phytoalexins (Bell, 1981). We have found that other phenolic substances such as vanillin and o-vanillin also promote haustoria, although not as actively as the constituents in gum tragacanth. Unfortunately, answers to the puzzle are incomplete. We have recently characterized the first root exudate stimulant from *Lespedeza serica* (Steffens

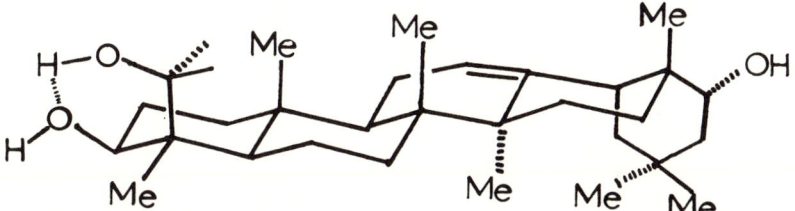

FIGURE 10. *Triterpene isolated from root exudate of* Lespedeza *active in promoting haustoria in* Agalinis purpurea.

et al., 1982, 1983). As shown in Fig. 10, it is a triterpene and quite different from the flavonoids. We find that it has low activity unless combined with other fractions of *Lespedeza* exudate. At present, we have not resolved this conflicting information. The *Lespedeza* isolations present problems because of the apparent presence of several constituents which collectively are very active, but individually have diminished activity. The constituents are also present at very low concentrations. We are not optimistic that it is presently feasible to identify all of them.

What we really want to know is if in plants like *Striga*, haustorial induction represents an evolved, receptor mediated recognition system. When water conditioned *Striga* seeds are placed on corn roots, they

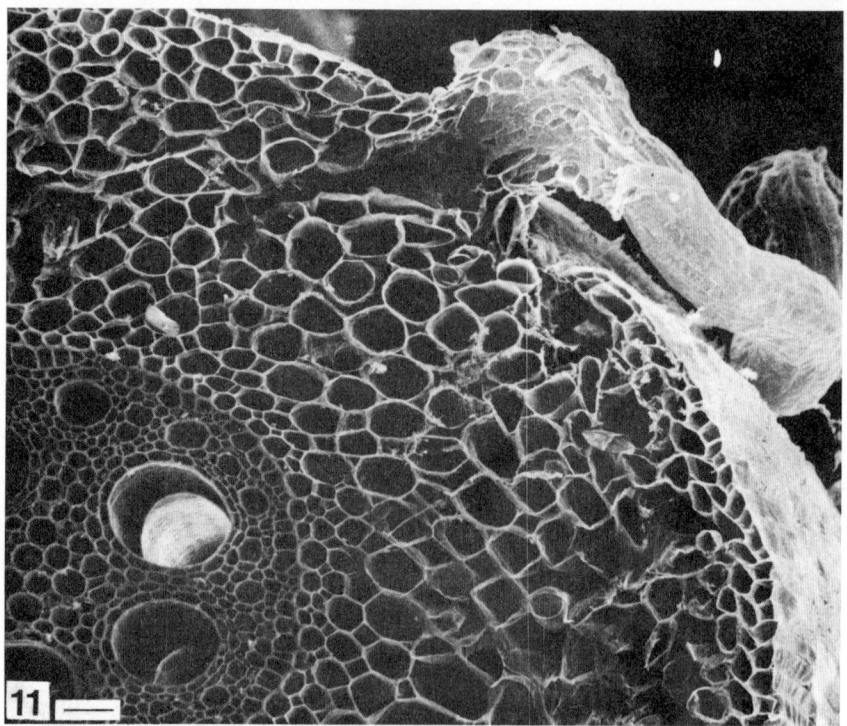

FIGURE 11. *Primary haustorium of 5 day old* Striga asiatica *seedling penetrating* Zea mays *root. (Scale 30 μm)*

germinate and develop a primary haustorium in 2-3 days (Fig. 11). This influence of host exudate in stimulating germination and haustoria was reported by Williams (1961). He also noted similar morphogenetic effects induced by light and kinetin. Recently, we have had success in separating a *Striga* germination stimulant and a haustorial inducing principle present in hydroponically grown corn. The stimulants were extractable with acetone from lyophilized exudate or extract and separable by silica gel chromatography. This work is continuing and supports the idea that *Striga* germination and haustorial initiation are developmental events regulated by separate, definable molecular signals. The first signal sets in motion germination and radicle extension. The root extension mode continues until the radicle is very close to the host root surface. A shift to the parasitic mode then occurs, characterized, as in *Agalinis*, by cessation of root elongation and in *Striga*, by the transformation of the primary root meristerm into the haustorium.

We know rather little about the regulation or the metabolic requirements for this shift from vegetative to parasitic mode. What we have observed in *Striga* is that the response for experimental induction of haustoria is confined to 3-5 days after germination. As shown in Fig. 12 the competence to respond to host signals is then lost even though radicle elongation may continue for 10-12 days.

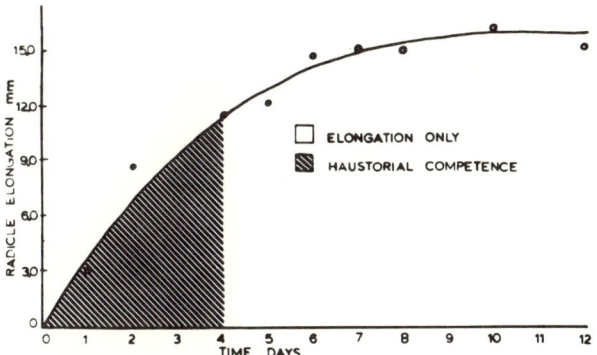

FIGURE 12.   *Profile of radicle elongation of* Striga asiatica *with haustorial competence limited to the first 4 days*

HAUSTORIAL ATTACHMENT

For some time, we have been interested in host-parasite relationships that influence the attachment of haustoria. This has principally been the work of one of my students, Vance Baird. The details will be published elsewhere, but I want to make briefly several points. First, once haustoria are induced, the attachment stage is totally indiscriminate. In field collections of *Agalinis*, the plant parasitizes a variety of herbaceous and woody plants (Riopel & Musselman, 1979). Self-attachments also occur. Indeed in laboratory experiments, we have found that haustoria will readily adhere to a variety of inanimate surfaces including glass and plastic. Second, the adhesion process involves specialized haustorial hairs quite different from the normal root hairs of *Agalinis*. SEM photos reveal the haustorial hair of *Agalinis* to have a papillate surface (Fig. 13) while the root hair is smooth (Fig. 14).

27

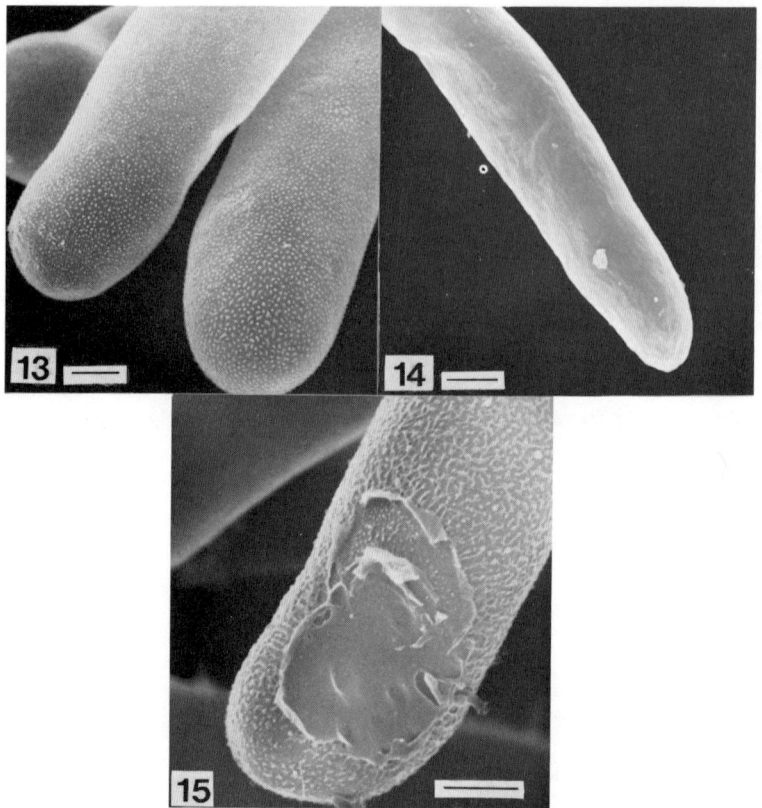

FIGURE 13. *Haustorial hair of* Agalinis purpurea *showing papillate surface.*
(Scale 20 μm)
FIGURE 14. *Normal root hair of* Agalinis purpurea *(Scale 20 μm)*
FIGURE 15. Agalinis *haustorial hair with attached host wall of* Lespedeza.
(Scale 20 μm)

Adhesive characteristics of the host-parasite bond can be demonstrated by removal of the haustorium which often results in tearing away a portion of the host cell surface (Fig. 15). Enzyme digestions and inorganic extractions show the papillae to be composed of hemicellulose and not pectin as might have been anticipated.

Third, time course studies have shown that haustoria can attach as early as six hours post-induction. By 12 hrs we found a 16% level of attachment and over 90% were attached by 36 hrs. The results of withholding the attachment substrate for specific periods after haustorial induction show that haustoria older than 60 hrs do not attach. This is surprising in view of the continued growth of *Agalinis* haustoria for periods of a month or more. New hairs are also initiated in older haustoria, but we notice that they frequently have fewer papillae.

The presence of papillae in other parasite species has not been studied by us. However, the general role of the haustorial hairs in attachment seems clear and not unlike the function ascribed in the previous report of clasping hairs for haustoria of *Pedicularis rostrato-spicata* (Weber and Uhlarz, 1976).

Attachment may also function in one other way. Haustoria that are not attached undergo little or no differentiation. Cells are parenchymatous with no organization, and little change occurs other than a gradual increase in cell number and the formation of haustorial hairs. Although there are no detailed studies that I am aware of, the process of attachment may convey a new message to the haustoria required for further differentiation. Certainly we know that vascular differentiation begins soon after attachment and during penetration of the host. There are also histochemical observations suggesting the synthesis of lytic enzymes which may begin during this period (Toth and Kuijt, 1977). We have looked for these enzymes in unattached haustoria of *Agalinis* and they are not present. From studies now in progress, we are hopeful that we can soon determine when biochemical processes of this kind are initiated and how interaction with the host and the attachment stage function in the regulation of these events.

SUMMARY

Early developmental stages in angiosperm root parasites are discussed under topics of seed germination, haustorial regulation, and haustorial attachment. The events of penetration and later host-parasite interactions are not considered.

The germination stages have been the most thoroughly studied in these plants. Most hemiparasites do not have specialized germination requirements. However, for germination, many of the holoparasitic species undergo a post-maturation stage and then require a water-conditioning treatment and an exposure to host root exudate. The biochemical mechanism activating germination in these plants is not understood. The most promising leads are suggested by the identification of strigol from nonhost cotton roots, a potent stimulator of germination in *Striga* and *Orobanche* species. Both in *Striga* and *Orobanche,* the requirement of a host root stimulant appears to be a reasonably effective host recognition step conserving the seeds of the parasite until a suitable host plant is nearby. Further studies are needed to evaluate fully the effectiveness of this stage in recognition and how it may operate. Laboratory studies show a wide range of promotive substances from nonhost root exudates as well as several synthetic compounds.

Although not true for all species, many root parasites do not initiate haustoria in the absence of host root exudate. Recently, *Agalinis purpurea* has been used as a model to examine the molecular basis for initiation. A variety of flavanoid molecules promote haustoria. Two flavonoids termed xenognosin A and B isolated from gum tragacanth have been studied in detail. Isomers of these compounds demonstrate considerable molecular specificity. Recently, a promotive triterpene, quite unrelated in structure to flavonoids has been identified from *Lespedeza* root exudate. In contrast to xenognosin, the triterpene has low promotive activity and may be functionally significant only in combination with other components of root exudate. In *Agalinis,* except for a few spontaneous haustoria, in the absence of a host plant of any kind, vegetative growth is sustained. The switch to the parasitic mode and the formation of haustoria comes only when a foreign root exudate impacts on the *Agalinis* root system. Following a period of reduced root elongation and the initiation of haustoria, the plant returns to normal primary root extension. In the continued presence of host exudate, more haustoria are added to the root system typically in lateral positions just proximal to the root tip.

The subsequent attachment stage is indiscriminate and will adhere to a variety of inanimate surfaces including glass and plastic. Experimental tests show that most *Agalinis* haustoria attach by 48 hrs post-induction. Haustoria continue to grow for several weeks but have diminished attachment competency.

The total contribution of the early developmental stages to host discrimination will not be clear until many other species are studied and especially those with more restrictive host preferences. In the meantime, it is evident that root parasites offer an excellent opportunity for studies on plant cell interactions.

ACKNOWLEDGEMENTS

The structural and chemical studies were carried out as a joint collaborative effort between this laboratory and that of Dr. David Lynn in the Department of Chemistry, University of Virginia. Special recognition goes to two Ph.D. students, Mr. John Steffens, who has been a major contributor to these chemical studies and to Mr. Vance Baird, who has been responsible for the attachment studies and all of the SEM work. Drs. Mike Thompson and Vinayak Kamat have also contributed significantly to these studies. This work was supported by USDA Competitive Research Grant 5901-0410-9-0257, USDA Cooperative Research Agreement 58-7B30-0-196, and Research Corporation.

LITERATURE CITED

Abu-Shakra, S., A. A. Mish, and A. R. Saghir. 1970. Germination of seed of branched broomrape (*Orobanche ramosa* L.) Hort. Res. *10:* 119-124.

Atsatt, P. R. 1977. The insect herbivore as a predictive model in parasite seed plant biology. Am. Nat. *111:* 579-586.

Atsatt, P. R., T. F. Hearn, R. T. Nelson, and R. T. Heineman. 1978. Chemical induction and repression of haustoria in *Orthocarpus purpurascens* (Scrophulariaceae). Ann. Bot. *42:* 1177-1184.

Ayabe, S. and T. Furuya. 1981. Biosynthesis of a retrochalcone, echinatin: a feeding study with advanced precursors. *Tetrahedron Lett. 22,* 2097-8.

Baird, Wm. Vance and J. L. Riopel. 1980. Studies on the life history of *Conopholis americana* (Orobancaceae). Bot. Soc. Am. Misc. series *159:* 9.

————— and —————. 1982. Initiation and early development of haustoria in *Agalinis purpurea.* Bot. Soc. Amer. Misc. series *162:* 9.

Beilin, I. G. 1968. Tsvetkovye Poluparazit i Phrazity. Moscow: Naukh. (in Russian).

Bell, A. A. 1981. Biochemical mechanisms of disease resistance. Ann. Rev. Plant Physiol. *32:* 21-81.

Botha, P. J. 1948. The parasitism of *Alectra vogelii* Benth. with special reference to the germination of its seeds. Jour. S. Afr. Bot. *14:* 63-80.

Botha, P. J. 1950. The germination of the seeds of angiospermous root parasites. I. The nature of the changes occurring during pre-exposure of the seed to *Alectra vogelii.* Benth. J. South Afr. Bot. *16:* 29-38.

Brown, R. and M. Edwards. 1944. The germination of the seed of *Striga lutea.* I. Host influence and the progress of germination. Ann. Bot., N.S. *8:* 131-148.

————— and M. Edwards. 1945. Effects of thiourea and allylthiourea on the germination of seed of *Striga lutea.* Nature (Lond) *155:* 455-456.

—————, A. W. Johnson, and G. J. Tyler. 1951. The *Striga* germination factor. II. Chromatographic purification of crude concentrates. Biochem. J. *50:* 596-600.

—————. 1965. The germination of angiospermous parasite seeds. Handburch der pflanzenphysiologie. *15:* 925-932. Springer-Verlag.

Cook, C. E., L. P. Whichard, B. Turner, M. E. Wall, and G. H. Egley. 1966. Germination of witchweed (*Striga lutea* Lour.): Isolation and properties of a potent stimulant. Sci. *154:* 1189-1190.

Cook, C. E., L. P. Whichard, M. E. Wall, G. H. Egley, P. Coggan, R. A. Luban, and A. T. McPhail. 1972. Germination stimulants. II. The structure of strigol—A potent seed germination stimulant for witchweed (*Striga lutea* Lour.) J. Amer. Chem. Soc. *94:* 6198-6199.

Chabrolin, C. 1938. Contribution a l'etude de la germination des graines de l'orobanche de la feve. Ann. Serv. Bot. Agron. Tunis. *14:* 91-145.

Davis, M., M. Pettett, D. B. Scanlon, V. Ferrito. 1977. Synthesis of some benzopyran derivatives related to the seed germination stimulant of *Orobanche crenata.* I. 3,3,A,4,9B-tetrahydro-2H-furo ((3,2-c)) ((1))benzopyrans. Aust. J. Chem. *30:* 228-92.

—————, —————, —————. 1978. Synthesis of some benzopyran derivatives related to the seed germination stimulant *Orobanche crenata* Forsk. 3,4,4,A,10б-tetrahydro-2H, 5H-pyrano [(2-C)] [(1)]benzopyrans. Aust. J. Chem. *31:* 1053-59.

Donini, B. 1959. Germination of *O. ramosa* independent of host's presence. Agric. Ital., *59:* 219-222.

Edwards, W. G. H. 1972. *Orobanche* and other plant parasite factors. In: *Phytochemical Ecology*. Ed. J. B. Harborne. Academic Press. pp. 235-248.

_____, R. P. Hiron, and A. I. Mallet. 1976. Aspects of the germination of *Orobanche crenata* seed. Z.Pflanzenphysiol. Bd. 80.S. 105-111.

Egley, G.H. 1972. Influence of the seed envelope and growth regulators upon seed dormancy in Witchweed (*Striga lutea* Lour). Ann. Bot. *36:* 755-770.

_____ and J. E. Dale. 1970. Ethylene, 2-chloroethylphosphonic acid and witchweed germination. Weed Sci. *18*(5): 586-589.

Eplee, R. E. 1981. *Striga's* status as a plant parasite in the United States. Plant Dis. *65*(12): 951-954.

French, R. C. and L. J. Sherman. 1976. Factors affecting dormancy, germination, and seedling development of *Aegenetia indica* L. (Orobanchaceae) Amer. J. Bot. *63*(5): 558-570.

Heinricher, E. 1898. Die grünen Halbachmarotzer. I. *Odontites, Euphrasia,* und *Orthantha.* Jahrb. Wiss. Bot. 31: 77-124.

Heinricher, E. 1901. Die grünen Halbachmarotzer. III. *Bartschia* und *Tozzia,* nebst Bemerkungen zur Frage nach der assimilatorischen Leistungsfahigkeit der grünen Halbachmarotzer. Jahrb. Wiss. Bot. 36: 665-753.

Hiron, R. W. P. 1973. An investigation into the processes involved in germination of *Orobanche crenata* using a new bio-assay technique. Proc. Eur. Weed Res. Coun. Symp. Parasitic Weds. Malta., pp. 76-88.

Hsiao, A. I., A. D. Worsham, and D. E. Moreland. 1981. Regulation of witchweed (*Striga asiatica*) conditioning and germination by dl-strigol. Weed Sci. *29:* 101-104.

Garas, N. A., C. M. Karssen, and J. Brunisma. 1974. Effects of growth regulating substances and root exudates on the seed germination of *Orobanche crenata* Forsk. Z. Pflanzenphysiol. *71:* 108-14.

Kadry, A. E. R. and H. Twefic. 1956. Seed germination of *Orobanche crenata*. Sv. bot. Tidskr. *50:* 270-286.

Kamat, V. S., Graden, D. W., Lynn, D. G., Steffens, J. C., and Riopel, J. L. 1982. A versatile total synthesis of xenognosin. Tetrahedron Lett., *23* (15): 1541-1544.

King, L. J. 1966. *Weeds of the World. Biology and Control.* New York: Interscience 526 pp.

Koch, L. 1887. Die Entwichlungsgeschichte der Orobanchen mit besonderen Berücksichtigung ihrer Beziehung zu den Kulturpflanzen. Ed. C. Winter, Heidelberg, 389.

Kuijt, J. 1969. *The Biology of Parasitic Flowering Plants.* Berkeley: Univ. Calif. Press. 246 pp.

Kuijt. J. 1977. Haustoria of phanerogamic parasites. Ann. Rev. Phytopathol. *15:* 91-118.

Kuijt, J. 1979. Host selection by parasitic angiosperms. Sym. Bot. Upsal. XXII: 4: 194-199. Uppsala.

Kusano, S. 1908. Further studies on *Aeginetia indica*. Beih. Bot. Centralb. *24*(1): 286-299.

Kust, C. A. 1966. A germination inhibitor in *Striga* seeds. Weeds *11:* 327-329.

Lihnell, D. 1942. Keimungsversuche mit *Pyrola*-Samen. Sym. Bot. Upsal. 6(3).

Lynn, D. G., J. C. Steffens, V. S. Kamat, D. W. Graden, J. Shabanowitz, and J. L. Riopel. 1981. Isolation and characterization of the first host recognition substance for parasitic angiosperms. J. Am. Chem. Soc. *103:* 1868-70.

Mann, W. F. Jr. and L. Musselman. 1981. Autotrophic growth of southern root parasites. Amer. Midland Naturalist *106*(1): 203-205.

Musselman, L. J. 1980. The biology of *Striga, Orobanche* and other root-parasitic weeds. Ann. Rev. Phytopath. *18:* 463-489.

Musselman, L. J. 1982. Parasitic weeds of arable land. In: *Biology* and *Ecology of Weeds.* Holzner, W. and N. Numata (eds.) Junk Pub. The Hague.

Nash, S. and S. Wilhelm. 1960. Stimulation of broomrape seed and germination. Phytopath. *50*(10): 772-774.

Okonkwo, S. N. C. 1966. Studies on *Striga senegalensis.* III. *In vitro* culture of seedlings. Establishment of cultures. Amer. J. Bot. *53*(7): 687-697.

_____. 1975. *In vitro* post-germination growth and development of embryos of *Alectra* (Scrophulariaceae). Physiol. Plant *34:* 378-383.

_____ and F. I. O. Nwoke. 1975. Bleach induced germination and breakage of dormancy of seeds of *Alectra vogelii.* Physiol. Plant. *35:* 175-180.

Pearson, H. H. W. 1912. On the Rooibloem (Isoma or Witchweed). Agr. Jour. Union S. Afr. *2*(3): 1-7.

Pavlista, A. D., A. D. Worsham, and D. E. Moreland. 1979. Witchweed seed germination. II. Stimulatory and inhibitory effects of Strigol, and GR7, and the effects of organic solvents. Proc. Second Symp. on Parasitic Weeds. North Carolina State University. Raleigh, N.C. pp. 228-237.

Privat, G. 1960. Recherches sur les phanerogames parasites ((Etude d' Orobanche hederae Duby) - These Doct. es Sc. Montpellier et in Ann. des. Sc. Nat. Bot. 12 e serie, 1960.

Riopel, J. L. 1979. Experimental studies on induction of haustoria in *Agalinis purpurea.* Proc. Second Symp. on Parasitic Weeds. North Carolina State University, Raleigh, N.C. pp. 165-173.

Riopel, J. L. and Musselman, J. 1979. Experimental initiation of haustoria in *Agalinis purpurea* (Scrophulariaceae). Amer. J. Bot. *66:* 570-575.

Saghir, A. R., M. Kurban, and B. Budayr. 1980. Studies on the control of *Orobanche* in Lebanon. Tropical Pest Mgr. *26*(1): 51-55.

Saunders, A. R. 1933. Studies in phanerogamic parasitism with particular reference to *Striga lutea* Lour. Dep. Agric. S. Afr. Bull. *128:* 1-57.

Spelce, D. L. and L. Musselman. 1981. *Orobanche minor* germination with Strigol and GR compounds. Z. Pflanzenphysiol. Bd. *104:* 281-283.

Steffens, J. C., D. G. Lynn, V. S. Kamat, and J. L. Riopel. 1982. Molecular specificity of haustorial induction in *Agalinis purpurea* (L) Raf. (Scrophulariaceae). Ann. Bot. *50:* 1-7.

_____, David G. Lynn and J. L. Riopel. 1982. Biochemistry of host recognition in *Agalinis purpurea.* Bot. Soc. Amer. Misc. series *162:* 24.

_____, J. L. Roark, D. G. Lynn, and J. L. Riopel. 1983. Host Recognition in Parasite Angiosperms: Use of Correlation Spectroscopy to Identify Long Range Coupling in an Haustorial Inducer. J. Am. Chem. Soc., In Press.

Subramanian, C. L. and A. R. Srinivasan. 1960. A review of the literature on phanerogamous parasites. Indian Council Agr. Res. Mon. 10. 96 pp.

Sunderland, N. 1960. Germination of the seeds of angiospermous root parasites. Brit. Ecol. Soc. Symp. (1959) *1:* 83-98.

Toth, R., J. Kuijt. 1977. Cytochemical localization of acid phosphatase in endophytic cells of the semiparasitic angiosperm *Comandra umbellata* (Santalaceae). Can. J. Bot. *55*(4): 470-475.

Vallance, K. B. 1950. Studies on the germination of the seeds of *Striga hermonthica.* I. The influence of moisture-treatment, stimulant-dilution and after-ripening on germination. Ann. Bot. N.S. *14:* 347-363.

_____. 1951. III. On the nature of pretreatment and after-ripening. Ann. Bot. N.S. *15:* 109-128.

33

Vaucher, J. P. 1823. Memoire sur la germination des Orobanches. Memoires du Museum d'Histoire Naturelle. *10:* 261.

Weber, H., H. Uhlarz. 1976. Die Kontaktaufnahme parasitischer Rachenblütler mit den Wurzeln ihrer Wirtspflanzen. Naturwiss. *6:* 296.

Williams, C. N. 1961. Growth and morphogenesis of *Striga seedlings.* Nature *189:* 378.

Worsham, A. D., D. E. Moreland, and G. C. Klingman. 1959. Stimulation of *Striga asiatica* (witchweed) seed germination by 6-(substituted) purines. Sci. *130:* 1654.

_____, G. C. Klingman, and D. E. Moreland. 1962. Promotion of germination of *Striga asiatica* seed by coumarin derivatives and effects on seedling development. Nature *195:* 199.

Yoshikawa, F., A. D. Worsham, D. E. Moreland, and R. E. Eplee. 1978. Biochemical requirements for seed germination and shoot development of witchweed (*Striga asiatica*). Weed Sci. *26:* 119-122.

III

Structural Responses of Plant Cells to Infection

M. G. Smart and J. R. Aist

Department of Plant Pathology, Cornell University
Ithaca, N.Y., 14853-0331

INTRODUCTION

At first glance, the responses of plant cells to infection appear bewildering in their variety. Plants may respond to infection by forming wall appositions, cicatrices, periderms or the more complex responses of the vascular wilt diseases. When examined in detail, these responses reduce to expressions of the same phenomenon, plant defense, albeit at different scales of intensity.

Basically, there is only a single response of plants to infection, a physiological one. The arbitrary separation of biochemical events from structural ones is useful in that it helps clarify our thinking about the interactions of pathogens and their hosts in resistance and disease.

In this chapter, structural responses include changes in the organization of the protoplasm, modifications to the structure of cell wall or developmental changes in tissue. Infection we define as the state of one organism (the parasite) becoming dependent on another (the host) after penetration of the host tissue. We cannot attempt to review the extensive literature exhaustively in the space available here. Nor has it been possible to discuss such topics as mycorrhizas, root nodulation in legumes, responses to nematodes or viral infections. We shall discuss cytoplasmic aggregation, papilla formation, cicatrices and wider responses of plants.

HIGHLY LOCALIZED RESPONSES

Protoplasmic Responses

The reaction of the protoplast to infection is often visually dramatic, especially when viewed in vivo. One commonly reported reaction is the aggregation of cytoplasm at the site of attempted penetration (Aist, 1976; Aist and Israel, 1977a; Aist and Israel, 1977b; Bushnell and Bergquist, 1975; Bushnell and Zeyen, 1976). When partially dissected coleoptiles of barley (Hordeum vulgare L.) were inoculated with conidia of the causal organism of powdery mildew (Erysiphe graminis (D.C.) Merat f. sp. hordei Em. Marchal), the aggregate formed about 10 hours later. Aggregates are quite common in plant infections (see review by Aist, 1976) but the timing of their appearance is, of course, quite variable. Cytoplasmic aggregation appears as a swirling, seething mass of granular cytoplasm (about 15 μm in diameter in barley) which can be swept away from its focus and dispersed, only to reform minutes later at the penetration site. It generally persists for several hours during the attempted penetration by the pathogen (Aist and Israel, 1977a; Aist and Israel, 1977b; Zeyen and Bushnell, 1979). Ultrastructurally, the aggregate contains large numbers of mitochondria and membrane bounded vesicles as well as dictyosomes and other cell components (Bushnell and Zeyen, 1976). It is not known if the concentration of any of these structures is greater in the aggregate than in the cytoplasm at large since no stereological analysis (Williams, 1977) has been reported.

As discussed by Aist (1976), the incitant of aggregate formation can be chemical. Observations that cells bordering those under stress can also form aggregates (Allen, 1923; Bushnell and Bergquist, 1975) are suggestive that chemicals are involved. Aggregrates can also be induced physically by wounding (Aist, 1976) so that the role of appressoria in induction cannot be dismissed.

Nuclei often respond to attack on the cell by migrating to the site of attempted penetration (Contreras and Boothroyd, 1975; Pappelis, et al., 1974; and Pearson, 1931). The movement of nuclei (and often the nuclei of neighboring cells) is of unknown significance.

Unfortunately, little is known of the mechanisms by which nuclei or cytoplasm are relocated to the point of attempted penetration. The principal difficulty is our poor understanding of cyclosis even in uninfected cells (Rebhun, 1972). Streaming appears to be based on microtubules (e.g. Allen, 1964) or, in other organisms, actin microfilaments (e.g. Allen, 1980). Preliminary experiments in this laboratory with the cytoplasmic aggregate induced in barley epidermal cells by E. graminis f. sp. hordei failed to show a fibrillar network (Smart, et al., 1982) even with the increased visibility of a video enhancement technique (Allen, et al., 1981).

Using Differential Interference Contrast optics, uninfected cells of the same coleoptile tissue show a network of dispersed, cytoplasmic fibrils. The fibrils appear to vary in thickness depending on the distance of transport; thicker fibrils are associated with saltatory organelle movements over several hundred micrometers (Smart, *et al.*, 1982). As far as we are aware, no other work directed at an understanding of aggregate formation has been attempted.

An aspect of the general protoplasmic response (independent of aggregation) which has received little attention is the changes in the numbers or distribution of cell organelles on infection. Increases in respiration rate are apparently ubiquitous in infected tissues (Beckman, 1980), and increases in the numbers of mitochondria or ribosomes are assumed to match those which occur in wounded tissues (Kahl, 1974). As in the case of the cytoplasmic aggregate, it is unfortunate that modern methods of stereological analysis (Williams, 1977) have not as yet been employed. These methods could provide valuable information on the spatial and temporal extent of the plant's response to infection.

Cell Wall Modifications

Papillae

The usual, although by no means inevitable, event which occurs during the existence of the cytoplasmic aggregate is the deposition of a papilla (Bushnell and Bergquist, 1975; Bushnell and Zeyen, 1976; Aist and Israel, 1977a; Aist and Israel, 1977b; Aist, 1976; Sherwood and Vance, 1976). Papillae are deposits of wall-like material between the cell wall proper and the plasmalemma. For many years they have been controversial candidates for a resistance mechanism (deBary, 1863; Smith, 1900; Akai, 1959; Bracker and Littlefield, 1973; Bushnell and Bergquist, 1975; Griffiths, 1971; Aist, 1976). In fact, much of the recent work has centered on this question and interested readers are referred to the above papers and reviews for discussion.

Papillae are generally 5 to 7 μm in diameter and, in cross-section, roughly hemispherical (Aist and Israel, 1977a; Aist and Israel, 1977b; Bushnell and Zeyen, 1976; Heath, 1972; Hachler and Hohl, 1982; Sherwood and Vance, 1976). Until the advent of the acoustic microscope, little was known of their physical properties. Bushnell (1971) removed a papilla from barley and found that it was hard. Recently, Israel and co-workers (1980) have employed acoustic microscopy of papillae in barley to show that resistant ones reflect more sound waves than the adjacent cell wall and are thus "more viscous, dense or elastic" than the surrounding wall.

The chemical composition of papillae is, like that of the cell wall itself, quite complex and still incompletely known. Since no methods have yet been reported for isolating papillae *en masse* from the cell wall there are only two available methods for their analysis. The first is to use mechanical isolation by micromanipulation (Kunoh, *et al.*, 1979). This approach has limited utility (for example in the determination of elemental composition) since only small numbers can be obtained. Using this method, Kunoh, Aist and Israel (unpublished) determined that papillae of barley induced by *E. graminis* f. sp. *hordei* contain moderate amounts of calcium and phosphorous but low amounts of silicon. This is in contrast to the papillae induced by the cowpea rust organism (*Uromyces phaseoli* (Pers.) Wint. var. *vignae* (Barcl.) Arth.) on bean (*Phaseolus vulgaris* L.) a non-host of this pathogen (Heath, 1979). Heath (1981) used EDAX of thin sections to characterize electron dense components of papillae and found them to contain high levels of silicon. That is, the composition of papillae in various host-pathogen interactions is not necessarily similar.

The second method of analysis, applicable to the organic fraction of papillae, is the use of histochemical means to determine composition. These investigations began almost a century ago when large amounts of callose (a β-1,3-glucan) were reported in papillae (Mangin, 1895). This compound is the most commonly reported constituent of papillae (Aist and Williams, 1971; Davison, 1968; Mercer, *et al.*, 1971; Nims, *et al.*, 1967; Sargant and Gay, 1977; Stanbridge, *et al.*, 1971; Sherwood and Vance, 1976). However, the usual methods employed for identifying callose are non-specific (Faulkner, *et al.*, 1973; Smith and McCully, 1978). The fluorescence of the fluorochrome in aniline blue has been used most often (Escrich, 1956). Smith and McCully (1978) suggest that this dye may be specific more for a physical conformation of the glucan than for β-1,3-glucans *per se*. In fact, it has been shown that aniline blue-positive material resulting from the incompatible response in pollen can have a low proportion of β-1,3-linkages (Vithanage, 1980). Other stains have been used to detect callose. These include lacmoid (resorcin blue) (Sherwood and Vance, 1976; Skou, 1982) and the related compound resorcinol blue (Sherwood and Vance, 1976). These stains are thought to be specific when differentiated with acid. Recently, Hinch and Clarke (1982) used highly purified β-1,3-glucanases in the first definitive proof of callose in papillae (see also Hachler and Hohl, 1982) in maize (*Zea mays* L.) induced by *Phytophthora cinnamomi* Rands, a non-pathogen of corn. Identification of callose or any other compound by its electron density is not reliable (O'Brien, 1972). Callose is thought to be impermeable to small ions at least *in vitro* (Escrich, 1975), and as such may play a role in the defense of plants by helping to sequester toxic compounds at the penetration sites (Aist, 1976).

Lignin is the second most commonly reported constitutent of papillae, however inexactly defined by histochemical tests (Vance, *et al.*, 1980;

Ride and Pearce, 1979; Mayama and Shishiyama, 1978). Papillae often autofluoresce (Aist and Israel, 1977a; Vance, *et al.,* 1980; Mayama and Shishiyama, 1978), and this fluorescence has usually been attributed to phenolics if not to lignin itself. The reliability of this assumption is not certain, and it is necessary to confirm the presence of phenolics by other means such as microspectrofluorimetry or staining reactions—preferably the former. Spectrofluorimetry has been used with some success (Mayama and Shishiyama, 1978), although the identity of the phenolics was not determined. There have been inconsistencies with staining reactions, at least in some systems (Mayama and Shishiyama, 1978; Aist and Israel, unpublished observations). In a most thorough histochemical investigation of papillae (Sherwood and Vance, 1976), lignin was detected by staining with toluidine blue O (O'Brien and McCully, 1964) chlorine-sulfite, and phloroglucinol-HCl (Jensen, 1962).

Tritiated phenylalanine fed to wheat (*Triticum aestivum* L.) leaves is preferentially incorporated into papillae and can be detected by autoradiography (Ride and Pearce, 1979). These authors coupled histochemical techniques with autoradiography to show lignin in the papillae (Ride and Pearce, 1979) induced by inappropriate pathogens. There seems to be little doubt that phenolics are present in papillae induced under these circumstances and their role as a defense mechanism may be substantial (see the review Vance, *et al.,* 1980). However, the role of phenolics in defense against appropriate pathogens of plants is less certain (Aist, 1976).

Other constituents of papillae which have been reported include suberin (Fellows, 1928), pectin (Ito, 1949), and cellulose (Sherwood and Vance, 1976). The function of all constituents of papillae, except possibly lignin (Vance and Sherwood, 1976), is unknown (Aist, 1976). It should be borne in mind that all these compounds may not always be present.

Ultrastructurally papillae are quite heterogenous (Bracker and Littlefield, 1973). After classical heavy metal staining, thin sections of papillae show a patchwork of areas of varying electron density interspersed with what appear to be cytoplasmic inclusions (Aist, 1976).

Haloes

There is often an area surrounding epidermal papillae in which a variety of staining reactions indicate that there has been an alteration in the cell wall. The exact nature of the alteration is uncertain. Using Interference microscopy, Russo and Pappelis (1981) have shown that the halo is an area of decreased mass per unit volume. They also showed decreases in the intensity of pectin staining, whereas cellulose staining increased in intensity. They therefore argue that the halo is an area of the host cell wall partially digested by enzymes of pathogen origin during penetration. Several laboratories have shown that the halo region fails to stain with reagents testing for cutin and cellulose (Sargant and Gay, 1977; McKeen, et al., 1968; Sargant, et al., 1973). Tests for reducing sugars are positive, suggesting localized enzymatic degradation (McKeen, et al., 1968).

Nevertheless, there is clear evidence for the infusion of molecules, particularly silicon, into the halo (Sargant, et al., 1973; Heath, 1979; Heath, 1981). Negative histochemical tests may be associated with decreased permeability in the halo due to this infusion (Sargant and Gay, 1977). The latter authors consider the halo as a sealant around the penetration peg, helping to control water loss from the infected cells.

GENERALIZED RESPONSES

Cicatrices and Periderms

The wider responses of plant cells to infection appear similar to the papilla response at least with respect to the changes in the cell wall. As will be described below, the cicatrix is usually lignified or suberized or both (Cunningham, 1928; Esau, 1977). The scale of the plant's response of wall modification would seem to depend on the extent of the irritation. Processes involved in the formation of the periderm, such as dedifferentiation, are of course unique in the response to infection.

Before discussing cicatrices and periderms any further, we should perhaps clarify what we mean by the two terms. According to Esau (1977), the cicatrix is the first layer of intact cells under a wound or at the edge of a necrosing area. This so-called "closing layer" becomes lignified or suberized. The periderm in many dicotyledons and gymnosperms (but not monocotyledons) forms a few cell layers beneath the cicatrix. This meristematic zone produces the phellem (cork), which also becomes suberized at maturity (Esau, 1977).

It is surprising that the literature on these types of tissue responses to wounding and infection is so limited. The major paper in the older literature is that by Cunningham (1928), and the principal modern works are those by Ride, Pearce and co-workers (1979; 1980; Maule and Ride, 1982), Mullick (1977 and references therein), and Shigo (1979, the Compartmentalization of Decay in Trees, CODIT, concept).

There are many groups of fungal pathogens which are able to penetrate into leaves but cannot develop extensively (e.g., *Coccomyces prunophorae* de Not. on *Prunus domestrica* L., *Cercospora beticola* Sacc. on *Beta vulgaris* L., Cunningham, 1928). However the majority of pathogens appear to fail for other, nonstructural reasons (Heath and Wood, 1969).

Cunningham's paper (1928) is the most extensive in the numbers of pathogen-suscept interactions examined. Because tissues were sampled at a single time after infection, this paper shows a suberized and lignified

40

cicatrix between the periderm and the healthy leaf tissue. No subsequent work has apparently attempted to clarify how a periderm can exist outside a cicatrix since the latter layer must presumably seal off the lesion (Esau, 1977; Botha, et al., 1982; Mullick, 1977). In sliced potato tubers at least, this is the case (Borchert and McChesney, 1973).

In trees, (Mullick, 1977) the initial cicatrix which forms, although neither suberized nor lignified, is impermeable and seals off the wound or infection. The products of periderm are usually suberized or lignified or both, but mature products of the phellogen form too late for them to be a primary determinant of disease resistance (Mullick, 1977). It is possible that the periderms which result from infections producing the shot-hole symptom are similarly unimportant. A role in the control of water loss from the perimeter of the lesion and the elimination of inoculum by ejection of infected tissue seem likely (Beckman, 1980).

The cicatrix proper may be involved in resistance; it simply has not been investigated thoroughly. The more obvious molecules which may be involved are suberin and lignin, but the existence of the non-suberized impermeable layer in trees (Mullick, 1977) should be borne in mind. The nature of the cell walls in this layer is unknown, nor for that matter is suberin itself well understood (see the review Kolattukudy, 1981). Suberin lamellae are probably impermeable to water (O'Brien and Carr, 1970; Botha, et al., 1982), but some pathogens are able to use them as a carbon source (Kolattukudy, 1981). Therefore, the relationship of this complex molecule to pathogenicity is unknown (but see Campbell, et al., 1980). Finally, localization of suberin by the dyes in the Sudan series may be confounded by infusion of phenolics or callose. Indeed, the phenolic nature of the suberin molecule is such that "lignin" stains may also react with it, in the absence of lignification.

The role of lignification in preventing fungal invasion is also being investigated (Ride and Pearce, 1979; Pearce and Ride, 1980; Vance, et al., 1980). Wheat leaves are generally lightly wounded and a non-pathogen (of wheat) introduced. Labelling studies suggest massive incorporation into both papillae and haloes, and a wider area which corresponds, to a cicatrix if it is extensive enough (Maule and Ride, 1982). There are good correlations between the lignification response and the cessation of fungal growth. It is still true, however, that in the absence of a time course study, the pathogen could have stopped growing for other reasons, and therefore lignification may be an unrelated event (Aist, 1976). Nevertheless, the lignified tissues are unavailable as a carbon source (Ride and Pearce, 1979).

The CODIT concept of disease resistance (Shigo, 1979) documents the responses of trees to infection over time period extending to years. Very little of this work details the structure of the lesion limiting cell walls which are responsible for the compartmentation (Tippett and Shigo,

1981; Pearce and Rutherford, 1981). However, the large amount of work on different tree genera shows clearly that invading microorganisms are contained within the cambium that is extant at the time of infection. The nature of the compartmentation may depend on cytoplasmic (that is, biochemical) factors rather than on structural ones. This aspect needs to be clarified.

Vascular Wilts

Plants respond in very complex and still incompletely understood ways when the vascular tissue is invaded (Beckman, 1980). The structural responses to the vascular wilt pathogens have been investigated most thoroughly, and it is these that we will discuss.

Usually a pathogen (e.g., *Verticillium albo-atrum* Reinke-Berth.) on tomato (*Lycopersicon esculentum* L.) is trapped at the first vessel ending in the file of xylem elements invaded (Beckman, 1980; Beckman, *et al.*, 1982). Subsequent reactions are the same regardless of the resistance or susceptibility of the plant; only the timing changes. First there is release of a gum (presumably derived from the xylem parenchyma cell wall) upon the release of pathogen cell wall degrading enzymes. Deposits of callose (Beckman, *et al.*, 1982; Meyer and Cote, 1968) appear in the neighboring xylem parenchyma cell walls on the wall closest to the vessel. The role of these deposits is unclear, but Beckman *et al.*, (1982) use them as a marker for resistance or susceptibility. Thirdly, there is apparently a change in the auxin levels in the tissues. The resultant local softening of the xylem parenchyma cell walls induces tylosis formation (Beckman, 1980; Beckman, *et al.*, 1982; Mace, 1978; Meyer and Cote, 1968; Talboys, 1958). If this response occurs above the level in the plant reached by the spores, the plant is resistant (Beckman, 1980).

CONCLUSION

From this brief overview of the cytology of structural plant responses to infection it should be clear that, although much excellent work has been done, large gaps still exist in our knowledge. Moreover, experimental proofs of many of the current postulates need to be completed. The structural interaction of plants and their parasites is an important aspect of general compatibility/incompatibility phenomena.

ACKNOWLEDGEMENTS

We wish to thank Christine Stockwell for help in the preparation of this chapter. One of us (MGS) wishes to acknowledge his debt to the kidney donor. This work was supported in part by NSF grant PCM 81-10822 and USDA grant 8100494 to Drs. J. R. Aist and H. W. Israel.

LITERATURE CITED

Aist, J. R. 1976. Papillae and related wound plugs. Ann. Rev. Phytopathol. *14*: 145-163.

Aist, J. R. and H. W. Israel. 1977a. Timing and significance of papilla formation during host penetration by *Olpidium brassicae*. Phytopathology 67:187-194.

Aist, J. R. and H. W. Israel. 1977b. Papilla formation: timing and significance during penetration of barley coleoptiles by *Erysiphe graminis hordei*. Phytopathology 67:455-461.

Aist, J. R. and P. H. Williams. 1971. The cytology and kinetics of cabbage root hair penetration by *Plasmodiophora brassicae*. Can. J. Bot. *49*:2023-2034.

Akai, S. 1959. Histology of defense in plants. *In* "Plant Pathology, An Advanced Treatise," Vol. I (J. G. Horsfall and A. E. Dimond, eds.), Academic Press, New York. pp. 391-434.

Allen, N. S. 1980. Cytoplasmic streaming and transport in the characean alga *Nitella*. Can. J. Bot. *58*: 785-796.

Allen, R. D. 1964. Cytoplasmic steaming and locomotion in marine foraminifera. In "Primitive Motile Systems in Cell Biology" (N. Kamiya and R. D. Allen, eds.), Academic Press, New York. pp. 407-432.

Allen, R. F. 1923. Cytological studies of infection of Baart, Kanred, and Mindum wheat by *Puccinia graminis tritici* forms III and XIX. J. Agric. Res. *26*:571-604.

Allen, R. D., N. S. Allen and J. L. Travis. 1981. Video- Enhanced Contrast, Differential Interference Contrast (AVEC-DIC) Microscopy: a new method capable of analyzing microtubule-related motility in the reticulopodial network of *Allogromia laticollaris*. Cell Motility *1*:291-302.

Beckman, C. H. 1980. Defenses triggered by the invader: physical defenses. In "Plant Disease, An Advanced Treatise" (J. G. Horsfall and E. B. Cowling, eds.), Vol. V, Academic Press, New York. pp. 225-245.

Beckman, C. H., W. C. Mueller, B. J. Tessier and N. A. Harrison. 1982. Recognition and callose deposition in response to vascular infection in fusarium wilt-resistant or susceptible tomato plants. Physiological Plant Pathology *20*: 1-10.

Borchert, R. and J. D. McChesney. 1973. Time course and localization of DNA synthesis during wound healing of potato tuber tissue. Dev. Biol. *35*:293-301.

Botha, C. E. J., R. F. Evert, R. H. M. Cross and D. M. Marshall. 1982. The suberin lamella, a possible barrier to water movement from the veins to the mesophyll of *Themeda triandra* Forsk. Protoplasma *112*: 1-8.

Bracker, C. E. and L. J. Littlefield. 1973. Structural concepts of host-pathogen interfaces. In "Fungal Pathogenicity and the Plant's Response" (R. J. Byrde and C. V. Cutting, eds.), Academic Press, London. pp. 159-318.

Bushnell, W. R. 1971. The haustorium of *Erysiphe graminis*: an experimental study by light microscopy. In "Morphological and Biochemical Events in Plant-Parasite Interaction" (S. Akai and S. Ouchi, eds.), The Phytopathological Society of Japan, Tokyo. pp. 229-254.

Bushnell, W. R. and S. E. Bergquist. 1975. Aggregation of host cytoplasm and the formation of papillae and haustoria in powdery mildew of barley. Phytopathology 65:310-318.

Bushnell, W. R. and R. J. Zeyen. 1976. Light and electron microscope studies of cytoplasmic aggregates formed in barley cells in response to *Erysiphe graminis*. Can. J. Bot. *54*:1647-1655.

Campbell, C. L., J. S. Huang and G. A. Payne. 1980. Defense at the Perimeter: the Outer Walls and the Gates. In "Plant Disease, An Advanced Treatise"

(J. G. Horsfall and E. B. Cowling, eds.), Vol. V, Academic Press, New York. pp. 103-120.

Contreras, M. R. and C. W. Boothroyd. 1975. Histological reactions and effects on position of epidermal nuclei in susceptible and resistant corn inoculated with *Helminthosporium maydis* Race T. Phytopathology 65:1075-1078.

Cunningham, H. S. 1928. A study of the histologic changes in leaves caused by certain leaf-spotting fungi. Phytopathology 18:717-752.

Davison, E. M. 1968. Cytochemistry and ultrastructure of hyphae and haustoria of *Peronospora parasitica* (Pers. ex. Fr.) Ann. Bot. (London) 32:613-621.

deBary, A. 1863. Recherches sur le developpement de quelques champignons parasites. Ann. Sci. Nat. Bot. Biol. Veg. 20:5-148.

Esau, K. 1977. "Anatomy of Seed Plants" John Wiley and Sons, New York, 2nd edition, 550 pages.

Escrich, W. 1975. Sealing systems in phloem. In "Transport in Plants. I. Phloem Transport." (M. H. Zimmermann and J. A. Milburn, eds.) Springer-Verlag, New York. pp. 39-56.

Escrich, W. 1956. Kallose. Ein kritisher sammelbericht. Protoplasma 47: 487-530.

Faulkner, G., W. C. Kimmins, and R. G. Brown. 1973. The use of fluorochromes for the identification of β(1-3) glucans. Can. J. Bot. 51:1503-1504.

Fellows, H. 1928. Some chemical and morphological phenomena attending infection of the wheat plant by *Ophiobolus graminis*. J. Agric. Res. 37:647-661.

Griffiths, D. A. 1971. The development of lignitubers in roots after infection by *Verticillium dahliae* Kleb. Can. J. Microbiol. 17:441-444.

Hachler, H. and H. R. Hohl. 1982. Histochemistry of papillae in potato tuber tissue infected with *Phytophthora infestans*. Botanica Helvetica 92:23-31.

Heath, M. 1972. Ultrastructure of host and nonhost reactions to cow-pea rust. Phytopathology. 62:27-38.

Heath, M. 1974. Light and electron microscope studies of the interactions of host and nonhost plants with cow-pea rust—*Uromyces phaseoli* var. *vignae*. Physiological Plant Pathology 4:403-414.

Heath, M. 1979. Partial characterization of the electron-opaque deposits formed in the nonhost plant, French bean, after cowpea rust infection. Physiological Plant Pathology 15:141-148.

Heath, M. 1981. Insoluble silicon in necrotic cowpea cells following infection with an incompatible isolate of cowpea rust fungus. Physiological Plant Pathology 19:273-276.

Heath, M. and R. K. S. Wood. 1969. Leaf spots induced by *Asochyta pisi* and *Mycospaerella pinodes*. Annals Bot. 132:657-670.

Hinch, J. M. and A. E. Clarke. 1982. Callose formation in *Zea mays* as a response to infection with *Phytophthora cinnamomi*. Physiological Plant Pathology 21:113-124.

Israel, H. W., R. G. Wilson, J. R. Aist and H. Kunoh. 1980. Cell wall appositions and plant disease resistance: Acoustic microscopy of papillae that block fungal ingress. Proc. Natl. Acad. Sci. USA 77:2046-2049.

Ito, K. 1949. Studies on "Murasaki-monpa" disease caused by *Helicobasidium mompa* Tanaka. Bull. Gov. For. Exp. St. (Japan) 43:1-126.

Jensen, W. A. 1962. "Botanical Histochemistry," W. H. Freeman and Co., San Francisco.

Kahl, G. 1974. Metabolism in plant storage tissue slices. Bot. Rev. 40:263-314.

Kolattukudy, P. E. 1981. Structure, biosynthesis and biodegradation of cutin and suberin. Ann. Rev. Plant Physiol. 32:539-567.

Kunoh, H., J. R. Aist and H. W. Israel. 1979. Microsurgical isolation of intact plant cell wall appositions for microanalysis. Can. J. Bot. *57*:1349-1353.

Mace, M. E. 1978. Contributions of tyloses and terpenoid aldehyde phytoalexins to *Verticillium* wilt resistance in cotton. Physiological Plant Pathology. *12*:1-11.

Mangin, L. 1895. Recherches sur les Peronosporees. Bull. Soc. Hist. Nat. Autun. 8:55-108.

Maule, A. J. and J. P. Ride. 1982 Ultrastructure and autoradiography of lignifying cells in wheat leaves wound-inoculated with *Botrytis cinerea*. Physiological Plant Pathology *20*:235-241.

Mayama, S. and J. Shishiyama. 1978. Localized accumulation of fluorescent and U.V.-absorbing compounds at penetration sites in barley leaves infected with *Erysiphe graminis hordei*. Physiological Plant Pathology *13*:347-354.

McKeen, W. E., R. Smith and P. H. Bhattacharya. 1968. Alterations of the host cell wall surrounding the infection peg of powdery mildew fungi. Can. J. Bot. *47*:701-706.

Mercer, P. C., R. K. S. Wood and A. D. Greenwood. 1971. Initial infection of *Phaseolus vulgaris* by *Colletotrichum lindemuthianum*. In "Ecology of Leaf Surface Micro-organisms" (T. F. Preece and C. H. Dickinson, eds.), Academic Press, New York, pp. 381-389.

Meyer, R. W. and W. A. Cote, Jr. 1968. Formation of the protective layer and its role in tylose development. Wood Sci. and Technology *2*:84-94.

Mullick, D. B. 1977. The non-specific nature of defense in bark and wood during wounding, insect and pathogen attack. In "Recent Advances in Phytochemistry" (F. A. Loewus and V. C. Runeckeles, eds.), Chapter 10, Plenum Publishing Corp., New York, pp. 396-441.

Nims, R. C., R. S. Halliwell and D. W. Rosberg. 1967. Wound healing in cultured tobacco cells following micro-injection. Protoplasma *64*:305-314.

O'Brien, T. P. 1972. The cytology of cell-wall formation in some eukaryotic cells. Bot. Rev. *38*:87-117.

O'Brien, T. P. and D. J. Carr. 1970. A suberized layer in the cell walls of the bundle sheath of grasses. Aust. J. Biol. Sci. *23*:275-287.

O'Brien, T. P., N. Feder, and M. E. McCully. 1964. Polychromatic staining of plant cell walls by Toluidine blue O. Protoplasma *59*:368-373.

Pappelis, A. J., G. A. Pappelis and F. B. Kulfinski. 1974. Nuclear orientation in onion epidermal cells in relation to wounding and infection. Phytopathology *64*:1010-1012.

Pearce, R. B. and J. P. Ride. 1980. Specificity of induction of the lignification response in wounded wheat leaves. Physiological Plant Pathology *16*:197-204.

Pearce, R. B. and J. Rutherford. 1981. A wound-associated suberized barrier to the spread of decay in the sapwood of oak (*Quercus robur* L.). Physiological Plant Pathology *19*:359-370.

Pearson, N. L. 1931. Parasitism of *Gibberella saubinetti* on corn seedlings. J. Agric. Res. *43*:569-596.

Rebhun, L. I. 1972. Polarized intracellular transport: Saltatory movements and cytoplasmic streaming. Int. Rev. Cytol. *32*:92-137.

Ride, J. P. 1980. The effect of induced lignification on the resistance of wheat cell walls to fungal degradation. Physiological Plant Pathology *16*:187-196.

Ride, J. P. and R. B. Pearce. 1979. Lignification and papilla formation at sites of attempted penetration of wheat leaves by non-pathogenic fungi. Physiological Plant Pathology *15*:79-92.

Russo, V. M. and A. J. Pappelis. 1981. Observations of *Colletotrichum circinans* f. *dematium* on *Allium cepa*: halo formation and penetration of epidermal cells. Physiological Plant Pathology *19*:127-136.

Sargant, J. A. and J. L. Gay. 1977. Barley epidermal apoplast structure and modification by powdery mildew contact. Physiological Plant Pathology 11:195-206.

Sargant, J. A., I. C. Tommerup and D. S. Ingram. 1973. The penetration of a susceptible lettuce variety by the downy mildew fungus Bremia lactucae Regel. Physiological Plant Pathology 3:231-239.

Sherwood, R. T. and C. P. Vance. 1976. Histochemistry of papillae formed in reed canary grass leaves in response to noninfecting pathogenic fungi. Phytopathology 66: 503-510.

Shigo, A. 1979. Tree decay: an expanded concept. Agric. Information Bull. #419 USDA Forest Service. 73 pp.

Skou, J. P. 1982. Callose formation responsible for the powdery mildew resistance in barley with genes in the ml-o locus. Phytopath. Z. 104:90-95.

Smart, M. G., N. S. Allen, C. A. Stockwell, J. R. Aist and H. W. Israel. 1982. Cytoplasmic responses to fungal attack; a fibrillar system related to organelle transport. (Abstr.). Phytopathology 72:933.

Smith, G. 1900. The haustoria of the Erysipheae. Bot. Gaz. 29:153-184.

Smith, M. M. and M. E. McCully. 1978. A critical evaluation of the specificity of aniline blue-induced fluorescence. Protoplasma 95:229-254.

Stanbridge, B., J. L. Gay and R. K. S. Wood. 1971. Gross and fine structural changes in Erysiphe graminis and barley before and during infection. In "Ecology of Leaf Surface Micro-organisms" (T. F. Preece and C. H. Dickinson, eds.), Academic Press, New York, pp. 367-379.

Talboys, P. W. 1958. Some mechanisms contributing to Verticillium-resistance in the hop root. Trans. Br. Mycol. Soc. 41:227-241.

Tippett, J. T. and A. L. Shigo. 1981. Barrier zone formation: a mechanism of tree defense against vascular pathogens. IAWA Bull. 2:163-168.

Vance, C. P., T. K. Kirk and R. T. Sherwood. 1980. Lignification as a mechanism of disease resistance. Ann. Rev. Phytopathol. 18:259-288.

Vance, C. P. and R. T. Sherwood. 1976. Regulation of lignin formation in reed canarygrass in relation to disease resistance. Plant Physiol. 57:915-919.

Vithanage, H. I. M. V., P. A. Gleeson and A. E. Clarke. 1980. The nature of callose produced during self-pollination in Secale cereale. Planta 148:498-505.

Williams, M. A. 1977. Quantitative methods in biology. In "Practical methods in electron microscopy" (A. M. Glauert, ed.), North Holland Publishing Co., Oxford.

Zeyen, R. J. and W. R. Bushnell. 1979. Papilla response of barley epidermal cells caused by Erysiphe graminis: rate and method of deposition determined by microcinematography and transmission electron microscopy. Can. J. Bot. 57:898-913.

IV

Physiological Responses of Plant Cells to Infection

A. A. Bell

USDA, National Cotton Pathology Research Laboratory
P.O. Drawer JF, College Station, TX 77841

INTRODUCTION

Diseased plants show many symptoms, such as growth abnormalities, chlorosis, wilting, water-soaking, and dark necrotic lesions. These symptoms represent the sum of various physiological responses of plant cells to metabolites either secreted by pathogens or released from pathogens by the action of host enzymes. Certain physiological responses, such as chlorosis and growth abnormalities, are characteristic of specific diseases, but apparently are not critical in determining the compatibility of the host and pathogen. These physiological responses will not be discussed in this review, but they have been reviewed elsewhere (Bell, 1977, 1981b).

The rapid necrosis of cells around infectious agents is a primary mechanism of plant incompatibility to pathogenic agents, especially when cell wall appositions, discussed in the previous chapter, fail to exclude pathogens from host cells. I refer to this mechanism of resistance as necrogenic resistance; it is also referred to as hypersensitive reaction (HR) and incompatibility reaction in plant pathology literature. Wilting may result from necrogenic resistance in xylem tissues, and watersoaking of leaves and stems often accompanies susceptible (compatible) reactions to bacterial pathogens. Thus, these symptoms will be considered because they reflect the incompatibility or compatibility of plants to pathogens.

Pathogenic organisms can be divided into three groups based on the nature or absence of necrogenic reactions induced in susceptible and

47

resistant hosts (Table 1). I have designated these pathogenic groups as biotrophs, transient biotrophs, and necrotrophs. Resistance (or incompatibility) to all groups is due to rapid necrogenic responses of host cells next to penetrating pathogens. In contrast, the groups have evolved diverse methods for causing susceptibility. Biotrophic pathogens, such as rust fungi, smut fungi, downy mildew fungi, mosaic viruses, and root-knot nematodes show complete compatibility with "susceptible" plants. These organisms reproduce with no appreciable death of host cells, and disease loss is due mostly to the drain of host metabolites into the biomass of the pathogen or into host tissues differentiated to support the pathogen. Transient biotrophs, such as *Phytophthora, Colletotrichum,* and *Cladosporium* species of fungi and bacterial pathogens, show a temporary period of compatibility with host cells, before necrogenic resistance is elicited. This period of compatibility generally lasts from one to several days, depending on the specific pathogen and host. Disease losses result both from the drain of host metabolites into the biomass of the pathogen and into the immune responses of the plant. Necrotrophic pathogens such as *Alternaria, Helminthosporium,* and *Botrytis* species of fungi, in contrast to biotrophs, do not show truly compatible relationships with either susceptible or resistant hosts. Rather, these pathogens secrete phytotoxins, including hydrolytic enzymes, that elicit extensive cell death and necrotic responses in the susceptible plant, but only limited necrotic responses in the resistant response. Necrosis in the susceptible plant has the same

TABLE 1. Classification of pathogens based on cellular reactions of the host.

Type of Pathogen	Cellular Reaction of Host	
	Susceptible (Compatible)[b]	Resistant[a] (Incompatible)
Biotroph	Remains alive around pathogen	Rapid death next to pathogen
Transient biotroph	Delayed death around pathogen	Rapid death next to pathogen
Necrotroph	Extensive death[c] beyond pathogen	Rapid death next to pathogen

[a]Resistant = Necrogenic Resistance = Hypersensitive Reaction (HR).
[b]For biotrophs and transient biotrophs only.
[c]This reaction can be considered as hyperincompatibility.

biochemical and physiological characteristics as the necrogenic response in the resistant plant (Bell, 1981a). Thus, susceptibility may be a form of hyperincompatibility, resulting from the plants resistance reaction to small molecules secreted by the pathogen. Some necrotrophs are also tolerant of necrogenic resistance responses and progress through tissues in spite of their occurrence.

FACTORS AFFECTING NECROGENIC RESISTANCE

The level of necrogenic resistance shown against a pathogen is a variable quality depending on many factors. At the highest level, a single cell may respond and completely stop the progress of the pathogen. As necrogenic resistance is weakened, either by delaying its onset or reducing its intensity, greater colonization by the pathogen occurs. Consequently, progressively more host cells become involved, resulting in progressively larger necrotic lesions, until finally systemic necrosis or blight symptoms may result.

Factors determining the degree of necrogenic resistance include: genotype, age, and specific tissue of the host (Bell, 1980; Milholland et al., 1981; Lazarovits et al., 1981; Stossel et al., 1981; Ward et al., 1981); genotype, age, and inoculum concentration of the pathogen (Bell and Mace, 1981; DeJager and Wesseling, 1981; Edwards and Agrios, 1981; Pilowsky, 1981); prior infections by other pathogens or damage by pests (Bell, 1981a, 1982; Cohen and Kuc, 1981; Hammerschmidt and Kuc, 1982; Hammerschmidt et al., 1982; McIntyre et al., 1981); and environmental components, such as temperature, light, nutrient levels, moisture, and agricultural chemicals (Anderbrhan and Wood, 1980; Bell, 1982; Deverall and McLeod, 1980; Giddix et al., 1981; Fraser and Loughlin, 1982; Prusky et al., 1981; Smith and Mansfield, 1981; Weststeijn, 1981). Multiple infections of a single plant stem by a transient biotroph may result in mixed susceptible (compatible) and necrogenic resistant (incompatible) responses at different infection loci (Lazarovits et al., 1981; Stossel et al., 1981; Ward et al., 1981). The fate of the whole stem then depends on the proportion of compatible loci as well as the total number of infection loci.

HISTOLOGY AND CYTOLOGY
OF NECROGENIC RESISTANCE

Necrogenic resistance may take on various appearances since a single cell or hundreds of cells may be involved, depending on the specific level of resistance. When only one or a few cells are involved, the resistant reaction can only be observed microscopically (Holliday et al., 1981). Various terms have been used to describe clusters of cells that show necrogenic resistance. The terms hypersensitive fleck and hypersensitive lesion are often used to describe necrogenic resistance reactions of leaves to fungi and bacteria, whereas similar reactions to viruses are referred to as local lesions. Special terms generally have not been devised to describe necrogenic reactions of internal tissues to pathogens, probably because these responses are not macroscopically visible. Yet, studies of necrogenic responses of vascular tissues to wilt fungi (Bell and Mace, 1981) and of protoxylem and protophloem to nematodes (Veech, 1982) indicate that necrogenic responses in these tissues are fully equivalent to the restricted lesions on leaves. This conclusion is supported by the fact that leaves from resistant potato cultivars show typical necrogenic responses to inoculations with the vascular pathogen *Verticillium albo-atrum* Reinke & Berth. (Hung and Whitney, 1982).

Cell death associated with necrogenic resistance may begin as early as 9 hr. after inoculation, but more commonly begins at 18-30 hr. and reaches a peak after a few days. Cell death in susceptible reactions to transient biotrophs occurs 1 to several days later than that in resistant reactions. Two examples illustrate the speed of necrogenic resistance. Bushnell (1981) observed collapse of resistant barley (*Hordeum vulgare* L.) cells at 18-26 hr. after inoculation with the biotrophic pathogen, *Erysiphe graminis* f. sp. *hordei* Em. Marchal. Cytoplasmic streaming halted 1-3 hr. before barley cells collapsed, and this halt was preceded by a 0.5 hr period in which cytoplasm accumulated around the fungal haustorium. No cell death occurs in reactions of susceptible cultivars. Holliday et al. (1981) studied the pattern of cell death in resistant soybeans (*Glycine max* (L.) Merr.) inoculated with the transient biotroph *Pseudomonas syringae* pv. *glycinea* (Coerper) Young et al. With high concentrations of the bacteria, small groups of one to five necrotic cells were observed in resistant cultivars 9 hr. after inoculation, whereas cell death in susceptible cultivars was not observed until 24 to 36 hr. With low bacterial concentrations, cell death began 24 hr. after inoculation in resistant cultivars and only after a few days in susceptible cultivars. Similar patterns of cell death have been described for necrogenic resistance involving many other host-pathogen combinations (Bell, 1981a).

Cells undergoing necrogenic resistance often show changes due to activated or accelerated secondary metabolism. Fluorescence under UV light is often seen because of accumulation of lignins, phytoalexins, and other secondary metabolites (Ersek et al., 1982; Holliday et al., 1981; Hargreaves, 1982; Kuck et al., 1981; Mansfield and Hutson, 1980). Electron microscopy usually reveals an accumulation of electron dense substances in necrogenic cells. These substances have most commonly been identified as polyphenolic substances (Obukowicz and Kennedy, 1981) but also may be silicates in cereals (Heath, 1981). Cells that have undergone necrogenic resistance usually are darkly pigmented because of the polyphenols and quinones that are synthesized and oxidized during the necrogenic process (Bell, 1981a; Pluck et al., 1981). Other anatomical changes in cells responding to pathogens are described in the previous chapter.

The development of necrogenic resistance generally causes a marked inhibition of pathogen growth in the diseased tissue. Pathogenic bacteria commonly develop populations of 10^6 cells or greater per gram in susceptible plants compared to about 10^3 in plants showing necrogenic resistance (Daub and Hagedorn, 1981; Smith and Mansfield, 1981). Likewise, hyphal development, haustorial development, and sporulation of pathogenic fungi are greatly restricted in resistant tissues (Bailey et al., 1980; Bailey and Rowell, 1980; Clifford and Roderick, 1981; Higgins, 1982; Keogh et al., 1980; Mansfield and Hutson, 1980; Pluck et al., 1981; Prusky et al., 1980; Rowell, 1981; Steinkamp et al., 1981; and Whitney and Mann, 1981). These observations indicate that a potent antibiotic environment develops during the necrogenic resistance response.

GENERAL METABOLIC CHANGES DURING NECROGENESIS

Cells undergoing necrogenic resistance show a number of metabolic changes that apparently are common to all host-pathogen combinations. These changes have been discussed in detail in four recent books (Bailey and Mansfield, 1982; Heitefuss and Williams, 1976; Horsfall and Cowling, 1980; Wood, 1982). The reader is directed to these books for most citations to original research in this area. Plants undergoing necrogenic resistance responses show metabolic changes similar to those in plants undergoing susceptible reactions to necrotrophs. Thus, the vast literature on metabolic responses to phytotoxins produced by necrotrophs (Durbin, 1981) may be useful in predicting metabolic responses associated with necrogenic resistance.

The following metabolic changes are generally observed during all forms of necrogenic reactions.

1. A general enhancement of metabolism occurs in cells undergoing necrogenesis and in adjoining cells. This is shown in electron micrographs by increased density of the endoplasmic reticulum, increased numbers of mitochondria, and increased numbers of Golgi bodies. Such changes, for example, are shown by broad bean leaves responding to *Botrytis* species (Mansfield and Richardson, 1981).

2. Membranes are altered. Invariably, there is a depolarization of membranes and subsequent leakage of electrolytes and small organic molecules from cells. An example is the leakage of electrolytes from cereal leaves undergoing necrogenic reactions to avirulent bacterial pathogens (Smith and Mansfield, 1981). There also is

51

an inhibition of K^+ uptake, H^+ extrusion, and K^+-activated ATPase at the plasmalemma during early stages of necrogenesis (Bell, 1981a). Finally, membranes invaginate and disrupt as cells become necrotic. Changes generally occur in the plasmalemma and tonoplast before they can be detected in mitochondrial or plastid membranes.

3. Respiratory rates are increased. An example is the marked increase of oxygen uptake of *Phaseolus lunatus* L. leaves during the formation of local lesions induced by the southern bean mosaic virus (Bell, 1964). The increase in respiration is largely due to enhanced activity of the pentose phosphate pathway of glucose breakdown. This pathway provides the essential sugars needed for nucleic acid and shikimic acid synthesis. It also produces the reduced cofactors needed for synthesis of secondary products.

4. Synthesis of RNA and protein is increased. Wagoner et al. (1982), for example, showed that 21 of 25 pea (*Pisum sativum* L.) proteins increased in concentration during responses to *Fusarium solani* (Mart.) Appel & Wr. They attributed the increased protein synthesis to increased activity of messenger RNA. Various inhibitors of protein synthesis and RNA synthesis block the necrogenic response in tissues and cause an increase in susceptibility to pathogens (Bell, 1981a).

5. Synthesis of plant hormones is enhanced. Cells undergoing the necrogenic response normally show a marked increase in ethylene evolution and in abscisic acid concentration (Bell, 1981b; Ketring and Melouk, 1982). Concentrations of IAA and cytokinins are often increased in surrounding tissues. The latter may contribute to the cellular hypertrophy and hyperplasia that commonly occurs next to necrotic cells. Increases in abscisic acid are apparently responsible for the stunted growth of plants resisting numerous infections (Whenham and Fraser, 1981).

6. Synthesis of cell wall substances, polyphenolic substances, and antibiotic substances is greatly increased. These changes are discussed in detail in the next section.

DEFENSE REACTIONS ASSOCIATED WITH NECROGENESIS

Several metabolic changes in cells showing necrogenic resistance confer direct resistance to the infectious agent. Some of these changes also may be responsible for killing the plant cell. These defense responses include: 1) the accumulation of antibiotics in the necrogenic cell and surrounding intercellular spaces, 2) the accumulation and oxidation of phenolic compounds derived from shikimic acid in both necrogenic cells and intercellular spaces, and the accumulation of nonoxidized phenolic compounds in surrounding cells, 3) the accumulation of enzymes that lyse microorganisms and of proteins that inhibit enzymes of pests, and 4) modification of cell walls to restrict or entrap pathogens. The first three responses will be discussed in more detail; response 4 is discussed in the previous chapter.

Accumulation of antibiotics

Antibiotics in necrotic cells originate from: 1) release of constitutive antibiotics contained within vacuoles or glands, 2) mixing of compartmentalized antibiotic precursors and enzymes, and 3) *de novo* synthesis of antibiotics not found in healthy tissue. The latter antibiotics are called phytoalexins and are the topic of a recent book (Bailey and Mansfield, 1982) that reviews the extensive literature on these compounds.

Hundreds of constitutive antibiotics and wound antibiotics are known (Bell, 1981a,b), but their roles in necrogenic resistance are unclear. These compounds normally are located only in specialized tissues, cells, or glands (Croteau and Winters, 1982; Mace et al., 1974) and histochemical reagents to detect them at the cellular level are usually lacking. Consequently, it is usually impossible to determine time-space relationships between these antibiotics and pathogens during necrogenic resistance responses.

The few studies that have been made indicate that constitutive antibiotics are released in early stages of necrogenic resistance. Oats (*Avena sativa* L.), for example, contain high levels of the constitutive avenacin antibiotics that appear to contribute to disease resistance (Holden, 1980). Fungal invasion of oat roots triggers release of avenacins from protoplasts simultaneously with electrolyte and nutrient leakage, which are characteristic of necrogenic resistance (Luning et al., 1978). Novacky (1972), likewise, showed that antibiotic terpenoid aldehydes were released from lysigenous glands of cotton (*Gossypium hirsutum* L.) into intercellular spaces during necrogenic resistant, but not susceptible, reactions to *Xanthomonas campestris* pv. *malvacearum* (Smith) Dye. Similar studies are needed to determine whether other constitutive antibiotics are released more rapidly during necrogenic resistant reactions than during compatible susceptible reactions.

The role of wound antibiotics in necrogenic resistance is even more difficult to evaluate because of the short half-lives of these compounds, the lack of specific histochemical reagents to detect them, and the diverse patterns of compartmentalization of the precursors and the enzymes that must be mixed to generate the antibiotics. For example, glucoside precursors in leaves of different plant species may occur uniquely in the epidermis (Thayer and Conn, 1981), uniquely in the mesophyll (Croteau and Winters, 1982), or in both epidermis and meso-phyll (Oba et al., 1981). Similarly, the glucosidase necessary to release the antibiotic may reside in the extracellular space of the cell containing the glucoside precursor, or it may reside in vacuoles of cells in the adjacent tissue. In the latter case, necrogenesis in both epidermal and adjoining mesophyll cells is necessary for formation of wound antibiotics. Indirect evidence indicates that wound antibiotics are important for resistance (Bell, 1977, 1981a,b), but more detailed studies are needed.

Extensive evidence indicates that the active synthesis of phytoalexins in response to attempted infection is a key part of the necrogenic resistance response in many plants. These studies have been reviewed in two recent books, "Phytoalexins" (Bailey and Mansfield, 1982) and "Active Defense Mechanisms in Plants" (Wood, 1982). Only a few recent references will be used in this article to illustrate the general principles that have emerged concerning phytoalexins.

Most known phytoalexins are unique natural products produced by only a limited number of plant genera or families. Most genera in a given plant family produce similar phytoalexins with only minor differences in structure. For example, all leguminous plants produce similar isoflavan phytoalexins, and all 32 Gossypium species and the seven other genera in the tribe Gossypieae produce similar sesquiterpenoid aldehyde phytoalexins. Most plants produce one or two major phytoalexins, plus several minor phytoalexins, all from the same class of chemicals (Burka et al., 1981; Weinstein et al., 1981; Uegaki et al., 1981). A few plants, however, produce phytoalexins from two different classes of chemicals. For example, peanut (Arachis hypogaea L.) produces both stilbene and isoflavan phytoalexins (Aguamah et al., 1981), and tomato (Lycopersicon esculentum Mill.) produces both polyacetylene and sesquiterpenoid phytoalexins (De Wit and Kodde, 1981a). Different tissues of the same plant may make different phytoalexins. For example, cotton leaves synthesize sesquiterpenoid naphthols and ketones (Essenberg et al., 1982), whereas xylem parenchymal cells synthesize sesquiterpenoid aldehydes (Bell and Stipanovic, 1978) in response to X. campestris pv. malvacearum. Further studies may reveal multiple phytoalexin systems and tissue specialization in additional species.

The synthesis of phytoalexins in response to biotrophic and transient-biotrophic pathogens invariably occurs coincidentally with necrogenesis. Thus, with biotrophic pathogens, a rapid, intense synthesis of phytoalexins occurs in cultivars showing necrogenic resistance, whereas little or no synthesis occurs in susceptible (compatible) cultivars (Cartwright and Russell, 1981; Mayama et al., 1982). With transient biotrophs, the quickness and intensity of phytoalexin synthesis is always greater in the cultivar showing necrogenic resistance than in the susceptible cultivar showing delayed necrosis (Bailey et al., 1980; Hutson and Smith, 1980; Holliday et al., 1981; Keogh et al., 1980; Langcake, 1981; Vaziri et al., 1981). Factors that enhance necrogenic resistance to transient biotrophs, such as light, temperature, age, and chemicals, also increase the quickness and intensity of phytoalexin synthesis (Anderbrhan and Wood, 1980; Cartwright and Langcake, 1980; Langcake, 1981; Lazarovits et al., 1981; Ward et al., 1981). In nearly all cases, microbial growth is restricted concurrent with phytoalexin synthesis.

Necrotrophic pathogens generally initiate early synthesis of phytoalexins in both susceptible or resistant plants. However, phytoalexin concentrations often increase more slowly in the susceptible plants. This slow accumulation is due to the conversion of the phytoalexin to a less toxic compound by the pathogen in several host-pathogen interactions (Arinze and Smith, 1980; Denny and Van Etten, 1981; Higgins and Ingham, 1981; Hutson and Mansfield, 1980; Smith et al., 1981). Tegtmeier and Van Etten (1982) conclusively showed that the ability to degrade pisatin is essential for high levels of virulence in *Nectria haematococca* Berk. and Br. to pea. In general, necrotrophs are much more tolerant of phytoalexins than are biotrophs or transient biotrophs, indicating that phytoalexin tolerance is an important determinant of their virulence.

Most phytoalexins are broad spectrum antibiotics that are toxic to animal cells, plant protoplasts, bacteria, and fungi. Gram positive bacteria are generally more sensitive to phytoalexins than gram negative bacteria (Gnanmanickam and Mansfield, 1981). Phytoalexins most toxic to fungi may be different than those most toxic to gram negative bacteria (Essenberg et al., 1982; Weinstein et al., 1981). Phytoalexins can be toxic to protoplasts of the same plant that produces them (see references in Bell, 1981a). Pisatin, a phytoalexin of pea, also inhibits stomatal opening in pea at $1 \times 10^{-4}M$ (Ayres, 1980). Thus, phytoalexins may contribute to the collapse of necrogenic cells and to other symptoms of diseased plants.

Polyphenols Derived from Shikimic Acid

Cells undergoing necrogenic resistance may show greatly enhanced synthesis of cis-dihydroxyphenols, such as protocatechuic acid, gallic acid, hydrolyzable tannin (galloyl glucose esters), shikimic and quinic acid esters of caffeic acid (e.g., chlorogenic acid), catechins, condensed tannins (3C-8C linked polymers of catechin) and various 3′, 4′-dihydroxyflavonoids. Obukowicz and Kennedy (1981) showed that increases in polyphenols could be seen in necrogenic resistant cells of tobacco within 10 hr. of inoculation with a bacterial pathogen, whereas no increases were seen in susceptible cells even after 30 hrs. Synthesis of lignin, a polymer of cinnamyl alcohols, may also increase greatly during necrogenesis (Bird and Ride, 1981; Hammerschmidt and Kuc, 1982; Maule and Ride, 1982). Living cells surrounding necrotic cells frequently show enhanced levels of lignin and suberin (a mixed polymer of lignin and fatty acid derivatives) in thickened cell walls that are thought to act as barriers to further infection (Davies et al., 1981; Glazener, 1982; Pearce and Rutherford, 1981; Pennypacker et al., 1981). Newly synthesized lignin also may be an essential part of cell wall appositions that successfully prevent penetration by fungal pathogens (Mansfield and Hutson, 1980).

Enzymes involved in the synthesis or oxidation of phenolic compounds also increase during necrogenic resistance. Synthesis of phenylalanine ammonia lyase, which converts phenylalanine to cinnamic acid, increased four to six fold within 4 hr. after inoculation of peas with necrotrophic pathogens (Loschke et al., 1981). Similar rapid increases in phenylalanine ammonia lyase activity have been observed in many other plants. Peroxidase enzymes polymerize cinnamyl alcohols into lignins, and catechins into condensed tannins. Peroxidase and polyphenoloxidase also oxidize cis-dihydroxyphenols to ortho-quinones, which are probably the major dark colored pigments in necrotic cells. Activities of both enzymes generally increase markedly during necrogenic resistance. For example, marked increases in polyphenoloxidase were seen in chloroplasts within 20 hr. after inoculation of resistant, but not susceptible, tobacco leaves with bacterial pathogens (Obukowicz and Kennedy, 1981). Similarly, Venere (1980) showed that peroxidase activity doubled within 12 hr. after inoculation in cotton leaves undergoing necrogenic resistance reactions to bacteria, whereas activity was unchanged in susceptible leaves. This indicates that increases in peroxidase activity occur early in the necrogenic resistance response. Treatments that enhance levels of necrogenic resistance also cause systemic increases of peroxidase activity throughout the plant (Hammerschmidt et al., 1982).

Few attempts have been made to determine whether cells producing phytoalexins also produce polyphenols and lignins. In cotton, at least, there appears to be cellular specialization for synthesis of either terpenoid aldehyde phytoalexins or condensed tannin polyphenols (Bell and Stipanovic, 1978). Thus, paravascular cells undergoing necrogenic reactions normally synthesize phytoalexins, whereas adjoining xylem ray cells synthesize condensed tannin. Likewise, epidermal cells of the root synthesize terpenoids, but underlying hypodermal cells synthesize tannins. The types of polyphenols synthesized by cotton leaves also change as a function of age. Infection of young leaves stimulates synthesis of flavonols (catechins and condensed tannins), whereas old leaves synthesize mostly flavonol glucosides,, such as isoquercitrin (Howell et al., 1976). Similar cellular specialization and age-related changes undoubtedly occur in other plants, but for the most part are unexplored.

Lytic Enzymes and Soluble Proteins

Plants produce 1,3-β-glucanase and chitinase which in combination can lyse fungal cells. Purified glucanase also digests glucans purified from fungi (Cline and Albersheim, 1981b). Fungal infections that cause necrosis may induce increases in 1,3-β-glucanase and chitinase activity in host tissues. In tobacco, the induction of these enzymes apparently is a unique response to fungi, since it does not occur with two nonfungal disorders causing necrosis (Edreva and Georgieva, 1980). In some cases, the speed and intensity of enzyme accumulation in response to fungal pathogens is directly related to the degree of necrogenic resistance (see references in Bell, 1981a), but in tomato infected with *V. albo-atrum*, glucanase activity increased more in susceptible than resistant plants (Pegg and Young, 1981; Young and Pegg, 1981). The cellular localization of increased enzyme activity is not known, and this response might not be part of necrogenic resistance. However, it is presumed that the enzymes are located near the pathogen, because fungal structures in diseased tissues often lyse and are not distinguishable after several days.

Diseased and wounded plants often show increases in various nonenzymic proteins. Some of these proteins inhibit pathogenic enzymes, such as proteinase (Ryan et al., 1981), while the biological significance of others is not known. DeWit and Bakker (1980) found major increases of two soluble proteins, as well as peroxidase, in tomato cultivars undergoing necrogenic resistance reactions to isolates of *Cladosporium fulvum* Cke. Much smaller increases occurred with susceptible reactions. Similar increases in "pathogenesis-related" proteins occur uniquely in young tobacco plants showing necrogenic resistance to virus infections (Pierpoint et al., 1981). However, the same proteins increase markedly in healthy older plants during flowering and are not consistently related to resistance (Fraser, 1981). Similar inconsistent results have been obtained for increases of hydroxyproline-rich proteins in cell walls. In two different host-pathogen systems, Esquerre-Tugaye et al. (1979) concluded that these are important for resistance, whereas Clark et al. (1981) concluded that they may be essential for compatible responses. Thus, the importance of lytic enzymes, enzyme inhibitors, and other proteins in necrogenic resistance is uncertain.

REGULATORS OF
NECROGENIC RESISTANCE

Compounds or treatments that "trigger" necrogenic resistance responses in plant cells are referred to as *elicitors*. These may be further divided into *biotic elicitors,* which are products from living organisms, and *abiotic elicitors,* which include physical treatments, such as cold shock and UV-irradiation, as well as various chemicals. Compounds or treatments that prevent or delay necrogenic resistance responses are called *suppressors*. Suppressors also may be of *biotic* or *abiotic* origin.

Biotic Elicitors

Live fungal cells, heat-killed conidia (Harding and Heale, 1981), culture filtrates (Ward et al., 1981), and exudates from germinating spores (Keogh et al., 1980; Mukhopadhyay and Sinha, 1980) elicit necrogenic resistance. Likewise, live cells, heat-killed cells, and extracellular polysaccharides of bacteria may elicit resistance responses (Bell and Stipanovic, 1978). Usually much higher concentrations of dead cells are required to cause the same effect as live cells.

DeWit and Kodde (1981b) isolated elicitors from culture filtrates of C. fulvum and identified them as glycoproteins. The glycoproteins, unlike live cells, caused necrogenic responses in both susceptible and resistant tomato fruits. Related glycopeptides apparently also occur on the surface of germ tubes of this fungus (Higgins, 1981) and might be more specific. Keen and Legrand (1980) have isolated surface glycoproteins from Phytophthora megasperma f. sp. glycinea (Drechs.) Kuan & Erwin, that specifically elicit necrogenic resistance in resistant cultivars of soybean.

Other molecules isolated from fungal cell walls that have been shown to act as nonspecific elicitors of necrogenic responses include glucans (Brown and Swinburne, 1981; Garas and Kuc, 1981; Hahlbrock et al., 1981; Yosikawa et al., 1981) and chitin and chitosans (Hadwiger and Loschke, 1981; Hadwiger et al., 1981; Pearce and Ride, 1982). Yoshikawa et al. (1981) showed that soybean tissues or extracts facilitated the release of glucan elicitors from mycelial walls of P. megasperma, indicating that host enzymes may release elicitors from pathogens. Similar observations have been made for release of host-specific phytotoxins (Larkin and Snowcroft, 1981). Also, El-Banoby et al. (1981) observed that extracellular polysaccharides from bacteria are broken down in intercellular spaces of resistant but not susceptible hosts. The products, however, were not examined for elicitor activity.

The most potent elicitor in culture filtrates of Rhizopus stolonifer is the enzyme polygalacturonase (Bruce and West, 1982; Lee and West, 1981a,b). The catalytic activity of the enzyme is necessary for it to function as an elicitor. When cell wall materials from castor bean (Ricinus communis L.) seedlings are treated with the enzyme, a heat-stable, water-soluble component that is highly active as an elicitor is obtained. This component apparently is a pectic fragment from the plant cell wall. Endogenous pectic elicitors also have been obtained from soybean cell walls by extraction with hot water or by partial acid hydrolysis (Hahn et al., 1981). Prolonged treatment with endopolygalacturonase destroys activity of the pectic elicitor, suggesting that the active molecule is a small polymer.

Most necrotrophic fungal pathogens and a few transient biotrophs exude compounds into the culture filtrate that cause necrosis and electrolyte leakage when administered to plant cells. These compounds are normally called phytotoxins because they disrupt cell membranes. Growing evidence, however, indicates that many of these compounds may be elicitors of necrogenic resistance responses.

The phytotoxins include simple molecules, such as cercosporin (Daub, 1982; Steinkamp et al., 1981), syringotoxin (Gonzalez et al., 1981), zinniol (Barash et al., 1981), dothistromin (Shain and Franich, 1981), and various unidentified molecules (Barash et al., 1981; Barrault et al., 1982; Dunkle and Wolpert, 1981), and complex molecules, such as peptidorhamnomannans (Nordin and Strobel, 1981; Russo et al., 1981), protein-lipopolysaccharides (PLP) (Nachimias et al., 1982) and other complex carbohydrates (Frantzen et al., 1982), that resemble the outer fungal or bacterial cell wall in composition. Some phytotoxins (at low concentrations) cause symptoms only in cultivars susceptible to the pathogen and therefore are called host-specific toxins. Bell (1981a,b) has summarized research findings that indicate that the necrogenic response to host-specific phytotoxins is the same as the necrogenic resistance response to biotrophs and transient biotrophs. In agreement with this concept, Dunkle and Wolpert (1981) found that electrolyte leakage (or membrane damage) alone could not account for the severe necrosis symptoms caused by the host-specific *Periconia* toxin in sorghum (*Sorghum vulgare* Pers.). Several polymeric compounds identified as nonspecific phytotoxins do act as elicitors (Bell et al., 1981a; Frantzen et al., 1982). The possibility that other necrogenic phytotoxins act as elicitors needs to be thoroughly explored because of the implications for disease control strategy and for agricultural applications of elicitors (Bell, 1981b).

Several attempts have been made to determine the minimum period for elicitors to trigger the necrogenic resistance reaction. The minimum period required for plant cells to "recognize" live bacteria is 2 to 3 hr. (Fett and Jones, 1982; Keen et al., 1981; Meadows and Stall, 1981). Early investigations (reviewed by Bell, 1981a) indicated that bacteria must attach to cell walls before the recognition process can begin. However, recent studies (Atkinson et al., 1981; Fett and Jones, 1982) show that attachment may not be important for recognition. Elicitor molecules from pathogens may act more quickly than the whole cells. Exposure of castor bean seedlings to fungal polygalacturonase for only 1 to 10 minutes before washing gave some elicitation of synthesis of the phytoalexin casbene. Hadwiger et al. (1981) also found that chitosan could be found in the protoplasm (especially in nuclei) within 15 minutes after application to cell surfaces. Thus, enzymes and wall components from pathogens once released may be able to trigger necrogenic resistance very rapidly.

Abiotic elicitors

Various chemical and physical treatments that disrupt cell membranes also elicit necrogenic resistance at critical concentrations or doses. In general, sulfhydryl binding reagents are excellent elicitors (Gustine, 1981). Thus, cupric ions are the best abiotic elicitors in cotton (Bell and Stipanovic, 1978), iodoacetate is the best abiotic elicitor in soybean (Keen et al., 1981) and mercuric ions are potent elicitors in several plants (Bell, 1981a). Detergents and organic solvents, such as chloroform (Bailey and Berthier, 1981), also are effective elicitors in some plants. The physical treatments most commonly used as elicitors are cold shock and UV-irradiation. Holding cotton seedlings at 5°C for 4 days causes a marked accumulation of terpenoid phytoalexins and necrosis in the elongation zone of the root. Physical elicitors are especially useful for studying biosynthesis of phytoalexins.

Biotic Suppressors

The ability of biotrophs and some transient biotrophs to maintain compatible relations with their host may depend on the synthesis and secretion of suppressor molecules. Host cells next to those infected by biotrophic pathogens frequently become susceptible to other microorganisms that they normally resist (Bell, 1981a). Virulent strains of pathogens also frequently produce antigens that crossreact with antibodies produced against similar antigens from susceptible hosts. These common antigens in pathogens and susceptible hosts may be necessary to delay or prevent recognition by the host. A common antigen in cotton apparently resides in the region of the plasmalemma (DeVay et al., 1981), which also seems to be a site of elicitor action. Crown-gall tissues from various plant species also have a common antigen in their cell walls which is missing from healthy plants (Galsky et al., 1981). Thus, the bacterial plasmid that is essential for virulence apparently induces some structural change in the plant cell wall or the plasmalemma of all host species. This change might be needed to suppress the necrogenic response in the host plant.

Extracellular polysaccharides (EPS) of bacteria have been shown to cause water soaking and suppress necrogenic responses in susceptible but not resistant plants (El-Banoby et al., 1980). In resistant plants the EPS is rapidly digested, apparently by host enzymes in intercellular spaces (El-Banoby et al., 1981). Rough mutants of bacteria that do not produce EPS invariably are avirulent regardless of bacterial species or host, further showing the importance of EPS as suppressors.

Extracellular polysaccharides or polysaccharides on the surface of fungal cells also may suppress necrogenic resistance. Doke and Tomiyama (1980) found that glucans isolated from various races of *Phytophthora infestans* (Mont.) d By. suppressed elicitor activity more effectively in susceptible than resistant potato (*Solanum tuberosum* L.) cultivars. N,N′-diacetyl-D-chitobiose, the hapten of potato lectin, also blocks necrogenic resistance in potato, suggesting that suppressors may attach to lectins on the plasmalemma (Nozue et al., 1980). This might block attachment or action of elicitors. A preparation of *P. infestans* elicitor was precipitated by purified potato lectin, but the elicitor did not lose appreciable activity (Garas and Kuc, 1981).

Abiotic Suppressors

Various antibiotics that block protein synthesis suppress necrogenic resistance responses (Bell, 1981a). Chemicals most used are cycloheximide and blasticidin S. High temperatures (30-40°C) during incubation, or temperature shock treatments (45-60°C for 1 to several minutes) prior to inoculation negate the hypersensitive reactions against viruses, bacteria, nematodes, or some fungi. Water-saturation of tissues also prevents or delays necrogenic resistance. These suppressor treatments further show that the necrogenic responses require active metabolism, including oxygen uptake, and are not a passive phenomenon due simply to rupture of cell membranes by metabolites of the pathogen.

ADVERSE EFFECTS OF NECROGENIC RESISTANCE

Although the necrogenic resistance response apparently is the normal immune response of plants to attempted invasion by pathogens, it also may cause undesirable effects in various plant interactions. These interactions include: 1) diseases caused by necrotrophic pathogens, particularly those involving host-specific phytotoxins, 2) interspecific incompatibilities such as pollen-stigma incompatibility, embryo abortion, or delayed lethal reactions in interspecific hybrid plants, 3) graft incompatibility, and 4) stunting by chemicals, saprophytic organisms, and environmental stresses. Mace and Bell (1981) showed that necrogenic resistance, including phytoalexin and tannin synthesis, occurs spontaneously in early stages of genetic lethal death in certain interspecific hybrids of *Gossypium*. They noted that this death is somewhat analagous to the autoimmune diseases, rheumatoid arthritis, lupus, and scleroderma, in animals. Cellular reactions in graft incompatibility (discussed in another chapter) also resemble closely those of necrogenic resistance to pathogens, suggesting that phytoalexins and phenolic polymers formed as a necrogenic resistance response may be involved. Stunting of plants is frequently accompanied by unexplained necrotic flecks on the root or in the xylem tissue. Environmental components may be eliciting extensive defensive reactions in tissues without causing a classical disease, in which a pathogen is found in the diseased tissue. Detailed knowledge of plant immune systems and their diverse functions is essential for dealing with various incompatibility phenomena in plant culture.

LITERATURE CITED

Aguamah, G. E., P. Langcake, D. P. Leworthy, J. A. Page, R. J. Pryce, and R. N. Strange. 1981. Two novel stilbene phytoalexins from *Arachis hypogaea*. Phytochem. 20:1381-1383.

Anderbrhan, T., and R. K. S. Wood. 1980. The effect of ultraviolet radiation on the reaction of *Phaseolus vulgaris* to species of *Colletotrichum*. Physiol. Plant Pathol. 17:105-110.

Arinze, A. E., and I. M. Smith. 1980. Antifungal furanoterpenoids of sweet potato in relation to pathogenic and nonpathogenic fungi. Physiol. Plant Pathol. 17:145-155.

Atkinson, M. M., J.-S. Huang, C. G. Van Dyke. 1981. Adsorption of pseudomonads to tobacco cell walls and its significance to bacterium-host interactions. Physiol. Plant Pathol. 18:1-5.

Ayres, P. G. 1980. Stomatal behaviour in mildewed pea leaves: solute potentials of the epidermis and effects of pisatin. Physiol. Plant Pathol. 17:157-165.

Bailey, J. A., and M. Berthier. 1981. Phytoalexin accumulation in chloroform-treated cotyledons of *Phaseolus vulgaris*. Phytochem. 20:187-188.

Bailey, J. A., and J. W. Mansfield. 1982. Phytoalexins. John Wiley and Sons, Halsted Press, New York.

Bailey, J. A., and P. M. Rowell. 1980. Viability of *Colletotrichum lindemuthianum* in hypersensitive cells of *Phaesolus vulgaris*. Physiol. Plant Pathol. 17:341-345.

Bailey, J. A., P. M. Rowell, and G. M. Arnold. 1980. The temporal relationship between host cell death, phytoalexin accumulation and fungal inhibition during hypersensitive reactions of *Phaseolus vulgaris* to *Colletotrichum lindemuthianum*. Physiol. Plant Pathol. 17:329-339.

Barash, I., H. Mor, D. Netzer, and Y. Kashman. 1981. Production of zinniol by *Alternaria dauci* and its phytotoxic effect on carrot. Physiol. Plant Pathol. 19:7-16.

Barash, I., G. Pupkin, L. Koren, G. Ben-Hayyim, and G. A. Strobel. 1981. A low molecular weight phytotoxin produced by *Phoma tracheiphila*, the cause of mal secco disease in citrus. Physiol. Plant Pathol. 19:17-29.

Barrault, G., B. Al-Ali, M. Petitprez, and L. Albertini. 1982. Contribution a l'etude de l'activite toxique de l'*Helminthosporium teres*, parasite de l'orge (*Hordeum vulgare*). Can. J. Bot. 60:330-339.

Bell, A. A. 1964. Respiratory metabolism of *Phaseolus vulgaris* infected with alfalfa mosaic and southern bean mosaic viruses. Phytopathology 54:914-922.

————. 1977. Plant pathology as influenced by allelopathy, pp. 64-99. IN The Role of Secondary Compounds in Plant Interactions (Allelopathy). USDA, ARS, Mississippi State.

————. 1980. The time sequence of defense, pp. 53-73. IN Horsfall, J. G. and E. B. Cowling (eds.). Plant Disease: An Advanced Treatise. Vol. V. How Plants Defend Themselves. Academic Press, New York.

————. 1981a. Biochemical mechanisms of disease resistance. Ann. Rev. Plant Physiol. 32:21-81.

————. 1981b. Morphogenic regulators in plant growth, disease development, and resistance. Proc. Plant Growth Regulators Working Group Meetings, St. Petersburg Beach, FL.

————. 1982. Plant pest interaction with environmental stress and breeding for pest resistance: Plant diseases, pp. 335-363. IN M. N. Christiansen and C. F. Lewis (eds.), Breeding Plants for Less Favorable Environments. John Wiley & Sons, Inc., New York.

Bell, A. A., and M. E. Mace. 1981. Biochemistry and physiology of resistance, pp. 431-486. IN M. E. Mace, A. A. Bell and C. H. Beckman (eds.), Fungal Wilt Diseases of Plants. Academic Press, New York.

Bell, A. A., and R. D. Stipanovic. 1978. Biochemistry of disease and pest resistance in cotton. Mycopathologia 65:91-106.

Bird, P. M., and J. P. Ride. 1981. The resistance of wheat to *Septoria nodorum*: fungal development in relation to host lignification. Physiol. Plant Pathol. 19: 289-299.

Brown, A. E., and T. R. Swinburne. 1981. Influence of iron and iron chelators on formation of progressive lesions by *Colletotrichum musae* on banana fruits. Trans. Br. Mycol. Soc. 77:119-124.

Bruce, R. J. and C. A. West. 1982. Elicitation of casbene synthetase activity in castor bean. Plant Physiol. 69:1181-1188.

Burka, L. T., L. J. Felice, and S. W. Jackson. 1981. 6-Oxodendrolasin, 6-hydroxydendrolasin, 9-oxofarnesol and 9-hydroxyfarnesol stress metabolites of the sweet potato. Phytochem. 20:647-652.

Bushnell, W. R. 1981. Incompatibility conditioned by the *Mla* gene in powdery mildew of barley: The halt in cytoplasmic streaming. Phytopathology 71:1062-1066.

Cartwright, D. W., P. Langcake, and J. P. Ride. 1980. Phytoalexin production in rice and its enhancement by a dichlorocyclopropane fungicide. Physiol. Plant Pathol. 17:259-267.

Cartwright, D. W., and G. E. Russell. 1981. Possible involvement of phytoalexins in durable resistance of winter wheat to yellow rust. Trans. Br. Mycol. Soc. 76:323-325.

Clarke, J. A., N. Lisker, D. T. A. Lamport, and A. H. Ellingboe. 1981. Hydroxyproline enhancement as a primary event in the successful development of *Erysiphe graminis* in wheat. Plant Physiol. 67:188-189.

Clifford, B. C., and H. W. Roderick. 1981. Detection of cryptic resistance of barley to *Puccinia hordei*. Trans. Br. Mycol. Soc. 76:17-24.

Cline, K. and P. Albersheim. 1981. Host-pathogen interactions. XVII. Hydrolysis of biologically active fungal glucans by enzymes isolated from soybean cells. Plant Physiol. 68:221-228.

Cohen, Y., and J. Kuc. 1981. Evaluation of systemic resistance to blue mold induced in tobacco leaves by prior stem inoculation with *Peronospora hyoscyami* f. sp. *tabacina*. Phytopathology 71:783-787.

Croteau, R., and J. N. Winters. 1982. Demonstration of the intercellular compartmentation of l-methone metabolism in peppermint (*Mentha piperita*) leaves. Plant Physiol. 69:975-977.

Daub, M. E. 1982. Peroxidation of tobacco membrane lipids by the photosensitizing toxin cercosporin. Plant Physiol. 69:1361-1364.

Daub, M. E., and D. J. Hagedorn. 1981. Epiphytic populations of *Pseudomonas syringae* on susceptible and resistant bean lines. Phytopathology 71:547-550.

Davies, W. P., B. G. Lewis, and J. R. Day. 1981. Observations on infection of stored carrot roots by *Mycocentrospora acerina*. Trans. Br. Mycol. Soc. 77:139-151.

Denny, T. P. and H. D. Vanetten. 1981. Tolerance by *Nectria haematococca* VI of the chickpea (*Cicer arietinum*) phytoalexins medicarpin and maackiain. Physiol. Plant Pathol. 19:419-437.

De Jager, C. P., and J. B. M. Wesseling. 1981. Spontaneous mutations in cowpea mosaic virus overcoming resistance due to hypersensitivity in cowpea. Physiol. Plant Pathol. 19:347-358.

Devay, J. E., R. J. Wakeman, J. A. Kavanagh, and R. Charudattan. 1981. The tissue and cellular location of a major cross-reactive antigen shared by cotton and soil-borne fungal parasites. Physiol. Plant Pathol. 18:59-66.

Deverall, B. J. and S. McLeod. 1980. Responses of wheat cells around heat-inhibited rust mycelia and associated with the expression of the Lr20, Sr6, and Sr15 alleles for resistance. Physiol. Plant Pathol. 17:213-219.

DeWit, P. J. G. M., and J. Bakker. 1980. Differential changes in soluble tomato leaf proteins after inoculation with virulent and avirulent races of Cladosporium fulvum (syn. Fulvia fulva). Physiol. Plant Pathol. 17:121-130.

DeWit, P. J. G. M., and E. Kodde. 1981a. Induction of polyacetylenic phytoalexins in Lycopersicon esculentum after inoculation with Cladosporium fulvum (syn. Fulvia fulva). Physiol. Plant Pathol. 18:143-148.

_____. 1981b. Further characterization and cultivar-specificity of glycoprotein elicitors from culture filtrates and cell walls of Cladosporium fulvum (syn. Fulvia fulva). Physiol. Plant Pathol. 18:297-314.

Doke, N., and K. Tomiyama. 1980. Suppression of the hypersensitive response of potato tuber protoplasts to hyphal wall components by water soluble glucans isolated from Phytophthora infestans. Physiol. Plant Pathol. 16:177-186.

Dunkle, L. D., and T. J. Wolpert. 1981. Independence of milo disease symptoms and electrolyte leakage induced by the host specific toxin from Periconia circinata. Physiol. Plant Pathol. 18:315-323.

Durbin, R. D. 1981. Toxins in Plant Disease. Academic Press, New York.

Edreva, A. M. and I. D. Georgieva. 1980. Biochemical and histochemical investigations of α- and β-glucosidase activity in an infectious disease, a physiological disorder and in senescence of tobacco leaves. Physiol. Plant Pathol. 17:237-243.

Edwards, M. C., and G. N. Agrios. 1981. Initiation and development of systemic necrosis in relation to virus concentration in tobacco ringspot virus-infected cowpea. Phytopathology 71:7-11.

El-Banoby, F. E., K. Rudolph, and A. Huttermann. 1980. Biological and physical properties of an extracellular polysaccharide from Pseudomonas phaseolicola. Physiol. Plant Pathol. 17:291-301.

El-Banoby, F. E., K. Rudolph, and K. Mendgen. 1981. The fate of extracellular polysaccharide from Pseudomonas phaseolicola in leaves and leaf extracts from halo-blight susceptible and resistant bean plants (Phaseolus vulgaris L.). Physiol. Plant Pathol. 18:91-98.

Ersek, T., M. Holliday, and N. T. Keen. 1982. Association of hypersensitive host cell death and autofluorescence with a gene for resistance to Peronospora manshurica in soybean. Phytopathology 72:628-631.

Esquerre-Tugaye, M.-T., C. Lafitte, D. Mazau, A. Toppan, and A. Touze. 1979. Cell surfaces in plant-microorganism interactions. Plant Physiol. 64:320-326.

Essenberg. M., M. A. Doherty, B. K. Hamilton, V. T. Henning, E. C. Cover, S. J. McFaul, and W. M. Johnson. 1982. Identification and effects on Xanthomonas campestris pv. malvacearum of two phytoalexins from leaves and cotyledons of resistant cotton. Phytopathology 72:1349-1356.

Fett, W. F., and S. B. Jones. 1982. Role of bacterial immobilization in race-specific resistance of soybean to Pseudomonas syringae pv. glycinea. Phytopathology 72:488-492.

Frantzen, K. A., L. B. Johnson, and D. L. Stuteville. 1982. Partial characterization of phytotoxic polysaccharides produced in vitro by Colletotrichum trifolii. Phytopathology 72:568-573.

Frazer, R. S. S. 1981. Evidence for the occurrence of the "pathogenesis-related" proteins in leaves of healthy tobacco plants during flowering. Physiol. Plant Pathol. 19:69-76.

Frazer, R. S. S., and S. A. R. Loughlin. 1982. Effects of temperature on the *Tm-1* gene for resistance to tobacco mosaic virus in tomato. Physiol. Plant Pathol. 20:109-117.

Galsky, A. G., J. A. Scheppler, and M. S. Cranford. 1981. Crown-gall tumors possess tumor-specific antigenic sites on their cell walls. Plant Physiol. 67:1195-1197.

Garas, N. A., and J. Kuc. 1981. Potato lectin lyses zoospores of *Phytophthora infestans* and precipitates elicitors of terpenoid accumulation produced by the fungus. Physiol. Plant Pathol. 18:227-237.

Giddix, L. R., F. L. Lukezic, and E. J. Pell. 1981. Effect of light on bacteria-induced hypersensitivity in soybean. Phytopathology 71:111-115.

Glazener, J. A. 1982. Accumulation of phenolic compounds in cells and formation of lignin-like polymers in cell walls of young tomato fruits after inoculation with *Botrytis cinerea*. Physiol. Plant Pathol. 20:11-25.

Gnanmanickam, S. S., and J. W. Mansfield. 1981. Selective toxicity of wyerone and other phytoalexins to gram-positive bacteria. Phytochem. 20:997-1000.

Gonzalez, C. E., J. E. DeVay, and R. J. Wakeman. 1981. Syringotoxin: a phytotoxin unique to citrus isolates of *Pseudomonas syringae*. Physiol. Plant Pathol. 18:41-50.

Gustine, D. L. 1981. Evidence for sulfhydryl involvement in regulation of phytoalexin accumulation in *Trifolium repens* callus tissue cultures. Plant Physiol. 68:1323-1326.

Hadwiger, L. A., J. M. Beckman, and M. J. Adams. 1981. Localization of fungal components in the pea-*Fusarium* interaction detected immunochemically with anti-chitosan and anti-fungal cell wall antisera. Plant Physiol. 67:170-175.

Hadwiger, L. A., and D. C. Loschke. 1981. Molecular communication in host-parasite interactions: hexosamine polymers (chitosan) as regulator compounds in race-specific and other interactions. Phytopathology 71:756-762.

Hahlbrock, K., C. J. Lamb, C. Purwin, J. Ebel, E. Fautz, and E. Schafer. 1981. Rapid response of suspension-cultured parsley cells to the elicitor from *Phytophthora megasperma* var. *sojae*. Plant Physiol. 67:768-773.

Hahn, M. G., A. G. Darvill, and P. Albersheim. 1981. Host-pathogen interactions. XIX. The endogenous elicitor, a fragment of a plant cell wall polysaccharide that elicits phytoalexin accumulation in soybeans. Plant Physiol. 68:1161-1169.

Hammerschmidt, R., E. M. Nuckles, and J. Kuc. 1982. Association of enhanced peroxidase activity with induced systemic resistance of cucumber to *Colletotrichum lagenarium*. Physiol. Plant Pathol. 20:73-82.

Hammerschmidt, R., and J. Kuc. 1982. Lignification as a mechanism for induced systemic resistance in cucumber. Physiol. Plant Pathol. 20:61-71.

Harding, V. K., and J. B. Heale. 1981. The accumulation of inhibitory compounds in the induced resistance response of carrot root slices to *Botrytis cinerea*. Physiol. Plant Pathol. 18:7-15.

Hargreaves, J. A. 1982. The nature of the resistance of oat leaves to infection by *Pyrenophora teres*. Physiol. Plant Pathol. 20:165-171.

Heath, M. C. 1981. The suppression of the development of silicon-containing deposits in French bean leaves by exudates of the bean rust fungus and extracts from bean rust-infected tissue. Physiol. Plant Pathol. 18:149-155.

Heitefuss, R., and P. H. Williams. 1976. Physiological Plant Pathology. Encyclopedia of Plant Physiology. Vol. 4. Springer-Verlag.

Higgins, V. J. 1982. Response of tomato to leaf injection with conidia of virulent and avirulent races of *Cladsporium fulvum*. Physiol. Plant Pathol. 20:145-155.

Higgins, V. J., and J. L. Ingham. 1981. Demethylmedicarpin, a product formed from medicarpin by *Colletotrichum coccodes*. Phytopathology 71:800-803.

Holden, J. 1980. Relationship between pre-formed inhibitors in oats and infection by *Gaeumannomyces graminis* and *Phialophora radicicola*. Trans. Br. Mycol. Soc. 75:97-105.

Holliday, M. J., N. T. Keen, and M. Long. 1981. Cell death patterns and accumulation of fluorescent material in the hypersensitive response of soybean leaves to *Pseudomonas syringae* pv. *glycinea*. Physiol. Plant Pathol. 18:279-287.

Horsfall, J. G., and E. B. Cowling. 1980. Plant Disease: An Advanced Treatise. Vol. V. How Plants Defend Themselves. Academic Press, New York.

Howell, C. R., A. A. Bell, and R. D. Stipanovic. 1976. Effect of aging on flavonoid content and resistance of cotton leaves to *Verticillium* wilt. Physiol. Plant Pathol. 8:181-188.

Hung, K. S., and N. J. Whitney. 1982. Reaction of two potato cultivars to leaf infection by *Verticillium albo-atrum*. Can. J. Bot. 60:554-556.

Hutson, R. A., and J. W. Mansfield. 1980. A genetical approach to the analysis of mechanisms of pathogenicity in *Botrytis/Vicia faba* interactions. Physiol. Plant Pathol. 17:309-317.

Hutson, R. A., and I. M. Smith. 1980. Phytoalexins and tyloses in tomato cultivars infected with *Fusarium oxysporum* f.sp. *lycopersici* or *Verticillium albo-atrum*. Physiol. Plant Pathol. 17:245-257.

Keen, N. T., T. Ersek, M. Long, B. Bruegger, and M. Holliday. 1981. Inhibition of the hypersensitive reaction of soybean leaves to incompatible *Pseudomonas* spp. by blasticidin S, streptomycin or elevated temperature. Physiol. Plant Pathol. 18:325-337.

Keen, N. T., and M. Legrand. 1980. Surface glycoproteins: evidence that they may function as the race specific phytoalexin elicitors of *Phytophthora megasperma* f. sp. *glycinea*. Physiol. Plant Pathol. 17:175-192.

Keogh, R. C., B. J. Deverall, and S. McLeod. 1980. Comparison of histological and physiological responses to *Phakopsora pachyrhizi* in resistant and susceptible soybean. Trans. Br. Mycol. Soc. 74:329-333.

Ketring, D. L., and H. A. Melouk. 1982. Ethylene production and leaflet abscission of three peanut genotypes infected with *Cercospora arachidicola* Hori. Plant Physiol. 69:789-792.

Kuck, K. H., R. Tiburszy, G. Hanssler, and H. J. Reisener. 1981. Visualization of rust haustoria in wheat leaves by using fluorochromes. Physiol. Plant Pathol. 19:439-441.

Langcake, P. 1981. Disease resistance of *Vitis* spp. and the production of the stress metabolites resveratrol., ε-viniferin, α-viniferin, and pterostilbene. Physiol. Plant Pathol. 18:213-226.

Larkin, P. J., and W. R. Scowcroft. 1981. Eyespot disease of sugarcane. Plant Physiol. 67:408-414.

Lazarovits, G., P. Stossel, and E. W. B. Ward. 1981. Age-related changes in specificity and glyceolin production in the hypocotyl reaction of soybeans to *Phytophthora megasperma* var. *sojae*. Phytopathology 71:94-97.

Lee, S.-C., and C. A. West. 1981a. Polygalacturonase from *Rhizopus stolonifer*, an elicitor of casbene synthetase activity in castor bean (*Ricinus communis* L.) seedlings. Plant Physiol. 67:633-639.

————. 1981b. Properties of *Rhizopus stolonifer* polygalacturonase, an elicitor of casbene synthetase activity in castor bean (*Ricinus communis* L.) seedlings. Plant Physiol. 67:640-645.

Loschke, D. C., L. A. Hadwiger, J. Schroder, and K. Hahlbrock. 1981. Effects of light and of *Fusarium solani* on synthesis and activity of phenylalanine ammonia-lyase in peas. Plant Physiol. 68:680-685.

Luning, H., U. Waiyaki, and G. Schlosser. 1978. Role of saponins in antifungal resistance. VIII. Interactions *Avena sativa-Fusarium avenaceum*. Phytopathol. Z. 92:338-345.

McIntyre, J. L., J. A. Dodds, and J. D. Hare. 1981. Effects of localized infections of *Nicotiana tabacum* by tobacco mosaic virus on systemic resistance against diverse pathogens and an insect. Phytopathology 71:297-301.

Mace, M. E. and A. A. Bell. 1981. Flavonol and terpenoid aldehyde synthesis in tumors associated with genetic incompatibility in a *Gossypium hirsutum* x *G. gossypioides* hybrid. Can J. Bot. 59:951-955.

Mace, M. E., A. A. Bell, and R. D. Stipanovic. 1974. Histochemistry and isolation of gossypol and related terpenoids in roots of cotton seedlings. Phytopathology 64: 1297-1302.

Mansfield, J. W., and R. A. Hutson. 1980. Microscopical studies on fungal development and host responses in broad bean and tulip leaves inoculated with five species of *Botrytis*. Physiol. Plant Pathol. 17:131-144.

Mansfield, J. W., and A. Richardson. 1981. The ultrastructure of interactions between *Botrytis* species and broad bean leaves. Physiol. Plant Pathol. 19:41-48.

Maule, A. J., and J. P. Ride. 1982. Ultrastructure and autoradiography of lignifying cells in wheat leaves wound-inoculated with *Botrytis cinerea*. Physiol. Plant Pathol. 29:235-241.

Mayama, S., Y. Matsuura, H. Iida, and T. Tani. 1982. The role of avenalumin in the resistance of oat to crown rust, *Puccinia coronata* f.sp. *avenae*. Physiol. Plant Pathol. 20:189-199.

Meadows, M. E., and R. E. Stall. 1981. Different induction periods for hypersensitivity in pepper to *Xanthomonas vesicatoria* determined with antimicrobial agents. Phytopathology 71:1024-1027.

Milholland, R. D., J. Papadopoulou, and M. Daykin. 1981. Histopathology of *Peronospora tabacina* in systemically infected burley tobacco. Phytopathology 71:73-76.

Mukhopadhyay, S., and A. K. Sinha. 1980. Spore germination fluid as inducer of resistance in rice plants against brown spot disease. Trans. Br. Mycol. Soc. 74:69-72.

Nachimias, A., V. Buchner, and J. Krikun. 1982. Comparison of protein-lipopolysaccharide complexes produced by pathogenic and non-pathogenic strains of *Verticillium dahliae* Kleb. from potato. Physiol. Plant Pathol. 20:213-221.

Nordin, J. H., and G.A. Strobel. 1981. Structural and immunochemical studies on the phytotoxic peptidorhamnomannan of *Ceratocystis ulmi*. Plant Physiol. 67:1208-1213.

Novacky, A. 1972. Influence of *Xanthomonas malvacearum* infection on cotton pigment glands. Plant Disease Reptr. 56:765-767.

Nozue, M., K. Tomiyama, and N. Doke. 1980. Effect of N,N'-diacetyl-D-chitobiose, the potato-lectin hapten and other sugars on hypersensitive reaction of potato tuber cells infected by incompatible and compatible races of *Phytophthora infestans*. Physiol. Plant Pathol. 17:221-227.

Oba, K., E. E. Conn, H. Canut, and A. M. Boudet. 1981. Subcellular localization of 2-(β-D-glucosyloxy)-cinnamic acids and the related β-glucosidase in leaves of *Melilotus albus* Desr. Plant Physiol. 68:1359-1363.

Obukowicz, M. and G. S. Kennedy. 1981. Phenolic ultracytochemistry of tobacco cells undergoing the hypersensitive reaction to *Pseudomonas solanacearum*. Physiol. Plant Pathol. 18:339-344.

Pearce, R. B., and J. P. Ride. 1982. Chitin and related compounds as elicitors of the lignification response in wounded wheat leaves. Physiol. Plant Pathol. 20:119-123.

Pearce, R. B., and J. Rutherford. 1981. A wound-associated suberized barrier to the spread of decay in the sapwood of oak (*Quercus robur* L.). Physiol. Plant Pathol. 19:359-369.

Pegg, G. F., and D. H. Young. 1981. Changes in glycosidase activity and their relationship to fungal colonization during infection of tomato by *Verticillium albo-atrum*. Physiol. Plant Pathol. 19:371-382.

Pennypacker, B. W., C. M. Smith, R. S. Dickey, and P. E Nelson. 1981. Histopathology of a symptomless chrysanthemum cultivar infected by *Erwinia chrysanthemi* or *E. carotovora* subsp. *carotovora*. Phytopathology 71:141-148.

Pierpoint, W. S., N. P. Robinson, and M. B. Leason. 1981. The pathogenesis-related proteins of tobacco: their induction by viruses in intact plants and their induction by chemicals in detached leaves. Physiol. Plant Pathol. 19:85-97.

Pilowsky, M., R. Frankel, and S. Cohen. 1981. Studies of the variable reaction at high temperature of F_1 hybrid tomato plants resistant to tobacco mosaic virus. Phytopathology 71:319-323.

Pluck, D. J., R. L. Evans, and C. L. Flegg. 1981. Resistance in barley to *Selenophoma donacis*. Trans. Br. Mycol. Soc. 77:509-518.

Prusky, D., A. Dinoor, and B. Jacoby. 1980. The sequence of death of haustoria and host cells during the hypersensitive reaction of oat to crown rust. Physiol. Plant Pathol. 17:33-40.

———. 1981. The fungicide or heat induced hypersensitive reaction of oats to crown rust: relations between various treatments and infection type. Physiol. Plant Pathol. 18:181-186.

Rowell, J. B. 1981. Relation of postpenetration events in Idaed 59 wheat seedlings to low receptivity to infection by *Puccinia graminis* f.sp. *tritici*. Phytopathology 71:732-736.

Russo, P. S., F. D. Blum, J. D. Ipsen, Y. J. Abul-Hajj, and W. G. Miller. 1981. The solubility and surface activity of the *Ceratocystis ulmi* toxin ceratoulmin. Physiol. Plant Pathol. 19:113-126.

Ryan, C. A., P. Bishop, G. Pearce, A. G. Darvill, M. McNeil, and P. Albersheim. 1981. A sycamore cell wall polysaccharide and a chemically related tomato leaf polysaccharide possess similar proteinase inhibitor-inducing activities. Plant Physiol. 68:616-618.

Shain, L., and R. A. Franich. 1981. Induction of Dothistroma blight symptoms with dothistromin. Physiol. Plant Pathol. 19:49-55.

Smith, D. A., J. M. Harrer, and T. E. Cleveland. 1981. Simultaneous detoxification of phytoalexins by *Fusarium solani* f. sp. *phaseoli*. Phytopathology 71:1212-1215.

Smith, J. J., and J. W. Mansfield. 1981. Interactions between pseudomonads and leaves of oats, wheat and barley. Physiol. Plant Pathol. 18:345-356.

Steinkamp, M. P., S. S. Martin, L. L. Hoefert, and E. G. Ruppel. 1981. Ultrastructure of lesions produced in leaves of *Beta vulgaris* by cercosporin, a toxin from *Cercospora beticola*. Phytopathology 71:1272-1281.

Stossel, P., G. Lazarovits, and E. W. B. Ward. 1981. Electron microscope study of race-specific and age-related resistant and susceptible reactions of soybeans to *Phytophthora megasperma* var. *sojae*. Phytopathology 71:617-623.

Tegtmeier, K. J., and H. D. Vanetten. 1982. The role of pisatin tolerance and degradation in the virulence of *Nectria haematococca* on peas: A genetic analysis. Phytopathology 72:608-612.

Thayer, S. S., and E. E. Conn. 1981. Subcellular localization of dhurrin β-glucosidase and hydroxynitrile lyase in the mesophyll cells of sorghum leaf blades. Plant Physiol. 67:617-622.

Uegaki, R., T. Fujimori, S. Kubo, and K. Kata. 1981. Sesquiterpenoid stress compounds from *Nicotiana* species. Phytochem. 20:1567-1568.

Vaziri, A., N. T. Keen, and D. C. Erwin. 1981. Correlation of medicarpin production with resistance to *Phytophthora megasperma* f. sp. *medicaginis* in alfalfa seedlings. Phytopathology 71:1235-1238.

Veech, J. A. 1982. Phytoalexins and their role in the resistance of plants to nematodes. J. Nematol. 14:2-9.

Venere, R. J. 1980. Role of peroxidase in cotton resistant to bacterial blight. Plant Sci. Lett. 20:47-56.

Wagoner, W., D. C. Loschke, and L. A. Hadwiger. 1982. Two-dimensional electrophoretic analysis of *in vivo* and *in vitro* synthesis of proteins in peas inoculated with compatible and incompatible *Fusarium solani*. Physiol. Plant Pathol. 20:99-107.

Ward, E. W. B., P. Stossel, and G. Lazarovits. 1981. Similarities between age-related and race-specific resistance of soybean hypoctyls to *Phytophthora megasperma* var. *sojae*. Phytopathology 71:504-508.

Weinstein, L. I., M. G. Hahn, and P. Albersheim. 1981. Host pathogen interactions. XVIII. Isolation and biological activity of glycinol, a pterocarpan phytoalexin synthesized by soybeans. Plant Physiol. 68:358-363.

Weststeijn, E. A. 1981. Lesion growth and virus localization in leaves of *Nicotiana tabacum* cv. *Xanthi* nc. after inoculation with tobacco mosaic virus and incubation alternately at 22°C and 32°C. Physiol. Plant Pathol. 18:357-368.

Whenham, R. J., and R. S. S. Fraser. 1981. Effect of systemic and local-lesion-forming strains of tobacco mosaic virus on abscisic acid concentration in tobacco leaves: consequences for the control of leaf growth. Physiol. Plant Pathol. 18:267-278.

Whitney, E. D., and N. F. Mann. 1981. Effect of resistance on growth of *Cercospora beticola* race C2 on the leaf surface and within leaf tissue of sugar beet. Phytopathology 71:633-638.

Wood, R. K. S. 1982. Active Defense Mechanisms in Plants. Plenum Press, New York.

Yoshikawa, M., M. Matama, and H. Masago. 1981. Release of a soluble phytoalexin elicitor from mycelial walls of *Phytophthora megasperma* var. *sojae* by soybean tissues. Plant Physiol. 67:1032-1035.

Young, D. H., and G. F. Pegg. 1981. Purification and characterization of 1,3-β-glucan hydrolases from healthy and *Verticillium albo-atrum*-infected tomato plants. Physiol. Plant Pathol. 19:391-417.

V

Structural Aspects of Graft Development

M. E. McCully

Department of Biology, Carleton University
Ottawa, Canada K1S 5B6

INTRODUCTION

Considering the long history of the horticultural use of grafted plants (for a comprehensive account of this fascinating story see Daniel, 1927) and their continuing economic importance, it is astonishing how little the structural aspects of graft development have been studied, particularly by modern techniques of tissue preparation and microscopy. This paucity of study is even more surprising in light of the widespread use of grafts in experimental investigations of such phenomena as transmission of the flowering stimulus (Zeevaart, 1976), the induction of vascular differentiation (Camus, 1949; Sachs, 1981) and the effects of relative position on the differentiation and function of the vascular cambium (e.g., Thair and Steeves, 1976; Warren Wilson and Warren Wilson, 1981). Furthermore, just how little is known in detail of the structure of developing grafts seems not to have been much considered in the interpretation of the results of such experiments. All of these studies, for example, postulate the movement of various morphogenic messages across the graft union, movement which is generally implied to be polar (and thus presumably symplastic). Yet, as will be discussed below, there is no good structural evidence for symplastic continuity all along a graft union. Indeed, plasmodesmata do not cross the interface of successful autografts at any stage of their development (McIntyre, McCully and Stoddard, in preparation). Similarly, there is little basis for the firmly held opinion that the vascular cambia of stock and scion must be in contact before a union can be effected (see discussion in Stoddard and McCully, 1979).

More recently, the discovery of complex mechanisms of cellular recognition in plants accompanying such interactions as those between host and parasite, host and symbiont, and pollen and stigma have occasionally led to the assumption that similar mechanisms also operate during graft formation. Again, as will be discussed below, our present (and very incomplete) knowledge of the structural events which occur during graft development does not in itself justify this assumption.

This review will attempt to summarize current knowledge of the structural events which occur during graft formation, and will also examine these events individually to determine if they are unique to the grafting process. Emphasis will center on graft development in herbaceous plants, due to their relative simplicity.

STRUCTURAL EVENTS WHICH OCCUR DURING THE DEVELOPMENT OF A SUCCESSFUL GRAFT UNION

The grafting process in autografts and compatible heterographs of dicotyledonous stems includes seven different structural events which can be easily detected in hand sections of fresh material taken at various times during graft development (Figs. 1-6). These events (usually appearing in this order) are: (1) the formation of necrotic material from the walls and contents of cut cells of the stock and scion (and subsequent changes in this material as the graft develops); (2) the extension of living cells of all tissue types lying immediately beneath the cut cells into the space between stock and scion (Fig. 1); (3) subsequent division of these enlarged cells (and also of many underlying cells) to form a callus (Fig. 2) which invades space between the graft partners (Fig. 3); (4) cohesion of the graft partners with strength enough to prevent separation during sectioning (Fig. 3); (5) differentiation of 'wound' type tracheary elements (Figs. 4, 5) and sieve tubes from callus cells and underlying parenchyma cells; (6) differentiation of a vascular cambium from callus cells and/or underlying parenchyma cells that links the original cambia of the stock and scion; and (7) production of secondary xylem and phloem by the new cambium, resulting in extensive vascular continuity across the graft interface (Fig. 6).

The events mentioned above can also be observed during graft formation in dicotyledonous roots (Stoddard and McCully, 1979) and in the development of woody grafts. The extensive earlier literature on graft anatomy (for review, see Stoddard and McCully, 1979; Stoddard, 1981) describes various combinations of some of these events as components

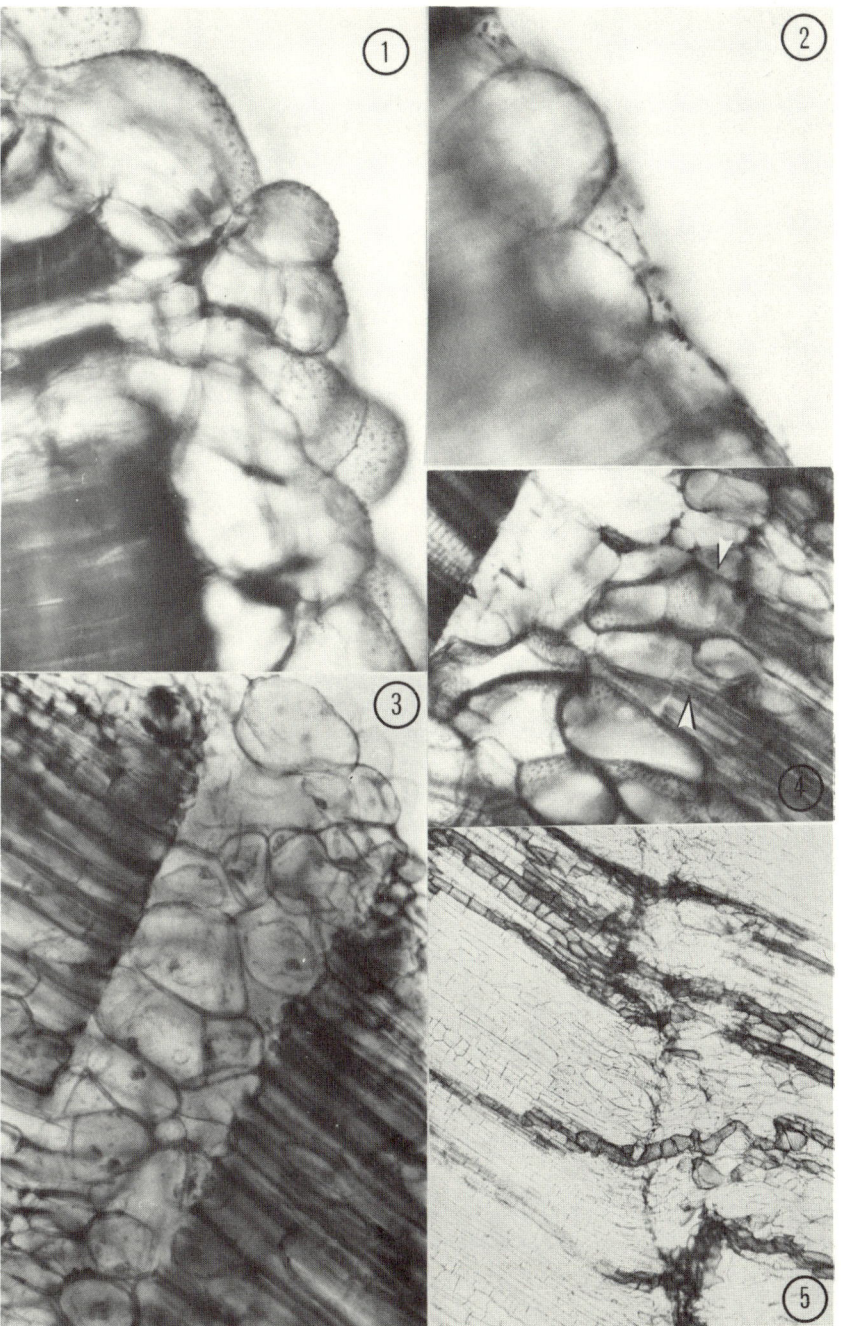

73

FIGURES 1 to 7 are longitudinal hand-cut sections of stem internode grafts. In Figs. 1 to 6 the stock is on the left side of the photomicrogrograph. In Figs. 7 to 9 the stock is at the bottom.

FIGURE 1. A portion of an autograft of tobacco (Nicotiana tabacum) 3 days after grafting. Stock and scion had cohered but were gently pulled apart. Callus has formed over the cut vascular regions of both stock (shown here) and scion. This callus is derived from expanded vascular parenchyma cells. The beaded structures on the surface of the outermost cells stain strongly bluish-green. The rest of the walls of the cells are strongly metachromatic pink. Toluidine blue staining. x 500.

FIGURE 2. Same section as shown in Fig. 1 but in the region of the pith parenchyma. The graft was also cohering in this region. No callus has formed here but a few of the underlying parenchyma cells have extended a little beyond the cut surface. These cells also have bluish-green staining beads on their surface. Toluidine blue staining. x 500.

FIGURE 3. A portion of a tobacco autograft 4½ days after grafting showing callus completely filling the graft space in the cortical parenchyma region. Some callus has pushed to the outside of the graft at the upper right. Callus is clearly derived from both graft partners, and it is impossible to tell where cells derived from each side have cohered. These grafts could not be pulled apart without tearing the cells. Callus cell walls are strongly metachromatic pink. The upper portion of cut parenchyma walls stain green. x 290.

FIGURE 4. Tobacco autograft at day 6. Some callus cells in the graft close to original vascular elements have differentiated to form wound xylem elements which bridge the graft. Some of the callus cells which invaded cut xylem element have differentiated to new xylem (arrows) there forming in effect lignified pegs. Secondary walls of the wound xylem element are stained green. Toluidine blue staining. x 290.

FIGURE 5. Coleus internode graft 6 days post-grafting. The tangential section was made just inside the cortical parenchyma. A few wound xylem elements bridge the graft. Section cleared in lactic acid and stained with aniline blue (Photomicrograph by F. L. Stoddard.) x 40.

FIGURE 6. Tomato (upper right) -tobacco heterograft (cleft graft) 6 weeks post-grafting. Portions of necrotic layer can still be seen along the graft line which runs between the arrow heads. Post-grafting cambial activity in both partners has produced considerable secondary xylem but has only recently bridged the graft. Most of the xylem across the graft is periclinally oriented wound xylem (x) derived from callus cells (c) which originally filled the space. Toluidine blue staining. x 150.

FIGURE 7. Coleus stem autograft 8 days post-grafting. The tangential section is cut just at the inside of the cambium (in plane of section). Many wound xylem elements bridge the graft line and some cambium-derived elements are apparent at the left side of the micrograph. Section cleared in lactic acid and stained with aniline blue. (Reprinted with permission of University of Chicago Press from Stoddard and McCully, 1980.) x 40.

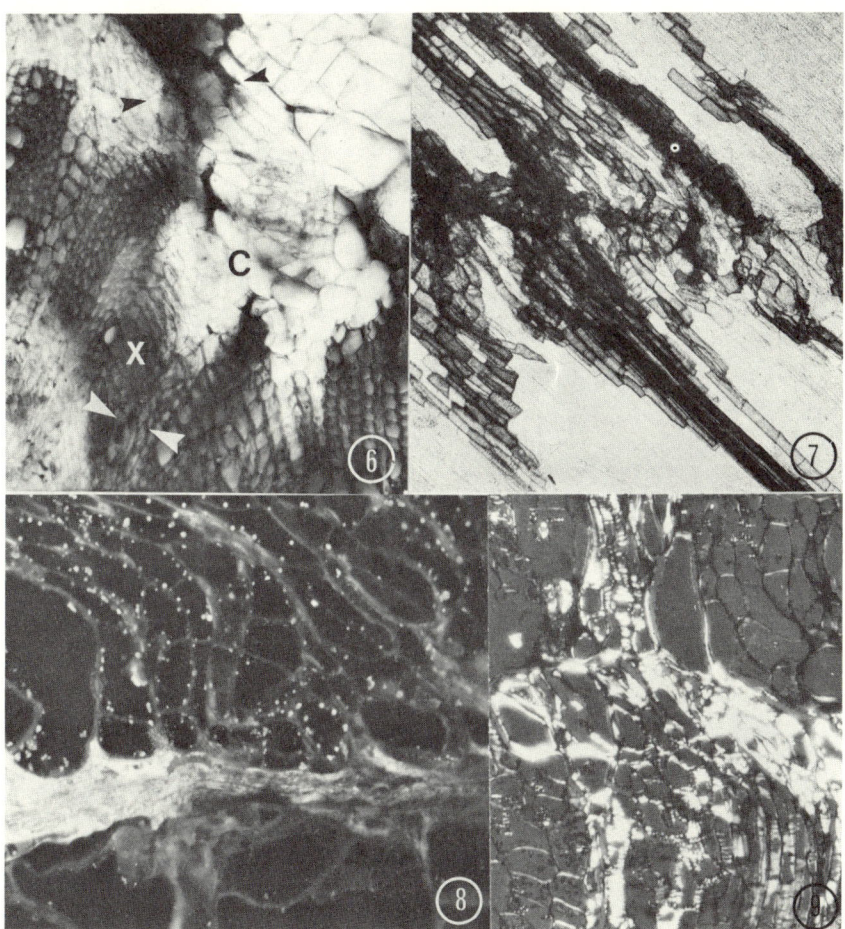

FIGURE 8. Stem graft of St. Augustine's grass (Stentotaphrum secundatum), 2 weeks post-grafting. The necrotic layer is very thick in places and no vascular tissue bridges the graft, though partners cohere strongly. Enlarged parenchyma cells are apparent on both sides of the graft. There is considerable starch accumulation in the scion. Non-specific staining by aniline blue. Fluorescence microscopy of glycol-methacrylate embedded tissue. (Photomicrograph by J. E. Bowerman.) x 330.

FIGURE 9. Stem graft as in Fig. 8 but 6 weeks post-grafting. Vascular tissue now bridges the graft and parenchyma cells at the graft line have developed thick secondary walls. Polarizing optics. (Photomicrograph by J. E. Bowerman.) x 250.

of graft formation. However, because many of these studies were not of a developmental nature, almost none recognized the complete sequence. Oddly, the event most often not discussed is cohesion of the partners!

A recent study of autografting in *Coleus* stems (Stoddard and McCully, 1980) provides the first comprehensive evidence that most of the above mentioned events occur independently in graft formation, and that they can be differentially manipulated by selective removal of portions of the stock and scion. When isolated internodes are grafted *in vitro,* a strong cohesion develops in the absence of cellular division, removal of, or change in the necrotic layer, or any vascular differentiation across the graft interface.

The different components of graft formation will now be considered in more detail.

Necrotic layer development—When the stock and scion are severed, cells at least one cell layer deep are killed at each surface. Necrosis may extend into living tissues well below the surface because of cutting of the tips of such elongated cells as fibers and vascular parenchyma. Initially, the necrotic material is probably composed of remnants of walls and contents of the cut cells. When graft partners are bound tightly together, the adjacent necrotic layers may be, from the beginning, collapsed and closely appressed. In more widely spaced regions, close appression of graft partners may subsequently occur because of expansion of underlying cells.

The necrotic layer approximately 5-10 μm thick is a feature of most published drawings and micrographs of young grafts of all types. Frequently, and particularly in earlier studies, staining methods are either not specified or are not specific for any one class of macromolecules. Thus, there is no clue to the histochemistry of the necrotic zone in these investigations (e.g., Figs. 2, 3, Muzik and La Rue, 1954; Figs. a, b, plate 2, Yeoman, et al., 1979; Fig. 63, Camus, 1949). However, where more specific staining reactions or optical methods have been used, the following properties of the necrotic region are noted: (1) It is strongly birefringent (e.g., Fig. 4; Stoddard and McCully, 1980); (2) It stains with reagents which react with lignins or polyphenols (e.g., acid fuchsin, toluidine blue, phloroglucinol); (3) The necrotic layer is autofluorescent during the early stages of graft formation (Stoddard and McCully, 1979); and (4) It shows weak but definite peroxidase activity (Fig. 2a, Deloire and Hébant, 1982).

The development of the necrotic layer has not been followed in enough cases to allow for generalizations, but evidence suggests that it is both complex and dynamic. For example, in almost all cases where good photomicrographs are presented (e.g., Muzik and La Rue, 1954), the necrotic layer appears too thick to be composed of dead cell remains only, even at an early stage in graft development. At later stages in the development of a grass graft (Fig. 8) the necrotic layer is markedly thickened (this feature has not been noted in dicotyledonous grafts).

Stoddard (1981), working with *Coleus* stem autografts, found two different materials apparently secreted into severed xylary elements close to the cut surface shortly after wounding. Both of these materials are autofluorescent, but the one which predominates in the stock stains green with toluidine blue while that in the scion stains deep violet. The two secretions also differ in a number of other staining properties. Some of these materials seem to be added to the necrotic zone, particularly in the vicinity of the cut xylem. In autografts of pea roots (Stoddard and McCully, 1979) and tobacco and tomato stem auto- and heterografts (McCully, unpublished), the initially greenish staining of the necrotic zone with toluidine blue is replaced after several days by metachromatic pink to purple staining. Over the same period of time the autofluorescence of the layer is much reduced.

In successful grafts where there is considerable disruption of the original cut surfaces by irregular tissue proliferations, much of the necrotic material disappears. However, some necrotic material often remains in pockets adjacent to and within the ends of the cut xylary elements and fibres (Figs. 20-25, Stoddard and McCully, 1979). Residue is also occasionally found between adjacent parenchyma cells. In regions where graft partners were originally in such firm contact that callus production was restricted, some original necrotic debris may remain tightly packed between walls of the underlying parenchyma cells, or all traces of such debris may be gone and the walls of the cells of the stock and scion so closely appressed that no 'middle lamella' is evident, even in high resolution micrographs. Two such quite different regions often adjoin (McIntyre, McCully, and Stoddard, in preparation).

While there is evidence of histochemical change and disintegration of much of the necrotic layer in successful grafts of entire dicot plants, the layer appears to persist in grafts between grasses (Fig. 8), possibly in grafts between buds and callus (Fig. 63, Camus, 1949), and in grafts of isolated *Coleus* internodes (Stoddard and McCully, 1980).

Cell enlargement and callus proliferation—Early in graft development (e.g., within 2 days in *Coleus* stem internode grafts; Stoddard, 1981) undamaged parenchyma cells which lie immediately under the cut surfaces of both the stock and scion begin to expand into the graft space. Often this expansion occurs by growth through the cavity which was the lumen of the adjacent cut cell. The lateral walls of the growing cell appear to adhere to the remains of the lateral walls of the cut cells, and the end wall bulges from the cut surface. Such bulges may become very large.

Expansion of parenchyma cells typically begins in the vascular parenchyma of both stock and scion (Stoddard and McCully, 1979; Stoddard, 1981), but can eventually occur anywhere along the graft line. However, it is usually less extensive in closely adhering regions of pith parenchyma.

Tylose development is also a prominent feature of the early stages of graft development. Living cells quickly intrude into adjacent xylary elements (e.g., Fig. 24, Stoddard and McCully, 1979). Copes (1969) illustrates beautifully the related event of resin canal epithelial cells expanding to fill a cut duct behind a graft union in Douglas fir. Frequently, a cut xylary element may be filled by expanding cells originating from the opposite partner. Thus, it is not always possible to unequivocally distinguish the origin of tylose-forming cells.

Expanding parenchyma cells may or may not subsequently divide to form callus (Fig. 1). Indeed, individual cells often expand greatly without dividing, especially in grafts of grasses (Figs. 8, 9). Typically, however, there is callus production. Although this may occur anywhere along the graft interface, it most commonly occurs in tissues originating in the vascular parenchyma and cortex (Fig. 1). Callus can intrude laterally along the graft space.

In addition to callus formation, there is always a lot of cellular division further removed from the cut surface, particularly in the cortical and vascular parenchyma regions. However, we have seen divisions in almost all cell types, including the endodermis (Stoddard and McCully, 1979; Stoddard, 1981). Expansion of surface and underlying callus cells must bring living cells and the remains of dead cells into firm contact at the graft interface.

The old idea that cellular division and callus production in grafts occur only from cambial origins was disproven early in the study of graft structure (see discussion in Saas, 1932; Crafts, 1934; Muzik and La Rue, 1952, 1954; Stoddard and McCully, 1979). This idea persists, however, almost certainly because of the importance of cambial contact between the stock and scion in establishing successful grafts of woody plants.

Differentiation of wound vascular tissue—In all cases where graft development has been closely examined, the first vascular tissues to bridge a graft are derived not from cambium, but from direct differentiation of parenchyma and callus cells to xylary elements and sieve tubes. This was recognized by such early workers as Simon (1930) and discussed by a number of subsequent workers (e.g., Crafts, 1934; Hayward and Went, 1939). However, the idea has persisted that vascular continuity is only a feature of the establishment of cambial activity across a graft. Wound xylem and phloem are always characterized by meandering strands of irregularly shaped cells (Figs. 5, 7). The elements are structurally identical to regenerating vascular elements described in other

78

wound situations (see Sachs, 1981). Strands always differentiate close to existing cut vascular strands. The course of development of the bridging strands has not been examined critically, but in pea root grafts (Stoddard and McCully, 1979) xylem is the first intact tissue. This differs from other wound situations in the same roots, where complete phloem strands are regnerated first (Robbertse and McCully, 1979).

The work of Stoddard and McCully (1980) with isolated *Coleus* internode grafts shows clearly the role played by scion leaves and buds in the promotion of wound vascular element differentiation across grafts.

Cohesion of graft partners—One of the most informative approaches to the study of graft development was that first used by Roberts and Brown (1961), by which the tensile strength of grafts could be determined over the course of their development. Subsequent studies by Lindsay, et al. (1974) and Moore (1983) have refined the methodology, but show essentially the same two phase initial development, with a slow increase in the tensile strength of the graft union over the first 3-4 days, followed by a period of 7-8 days in which there is a rapid rise in the tensile strength of the graft.

The correspondence of these time periods in two very different plant systems is remarkable. Unfortunately, the excellent data on the development of tensile strength is not well matched by a detailed knowledge of structural events occurring in all regions along the graft line in the systems studied, so that one cannot conclusively assess the relationship of these events to the data. Indeed, there is little consistency in the events that are cited in the 3 papers as correlating with the phases of the cohesive process. Roberts and Brown (1961), for example, found no meristematic activity in the graft zone over the first 4 days, while Lindsay, et al. (1974) report a large increase in meristematic activity over the same time period and propose that somehow this activity is ultimately necessary for the production of the 'glue' which sticks the stock and scion together. Lindsay, et al. (1974) correlate a correspondingly large increase in the number of tracheary elements in the graft zone with the second phase of adhesion, while Moore (1983) finds that the differentiation of wound xylary elements does not begin until almost the end of the period of rapid increase in graft strength.

Study of hand sections of developing tobacco stem autografts and tobacco-tomato heterografts (Figs. 1-4) shows a clear sequence of structural events which also could be linked to the same time course for adhesion. Stocks and scions cohere weakly by day 2, and can be gently separated without destroying cells up to 4 days after grafting. The separated surfaces are characterized by cellular protrusion and callus proliferation, and the surfaces of cells which face the graft space or have made weak connections with opposing cells are covered with small beaded structures (Fig. 1) which give a polyphenol-type staining reaction with toluidine blue (green to blue-green, in contrast to strong pink metachromasy for the remainder of these wall surfaces). By 4 days after grafting there is a considerable amount of interdigitation of the expanding callus masses, and some of these cells invade the cut ends of xylary elements and fibers of the opposite partner, presumably thus effecting a considerable increase in the tensile strength of the union. Also, by 4 days after grafting there is a first appearance of fully differentiated wound xylary strands across the graft (Fig. 4). Their number greatly increases over the next few days (as also occurs in *Coleus* stem grafts; Stoddard and McCully, 1981; Figs. 4, 5, 7). These lignified strands must increase the tensile strength of the graft union, especially where they have differentiated in continuity with newly formed xylem of the original vascular strands. Often the tyloses formed by callus invasion of cut xylary elements form the ends of the wound xylary strands (Fig. 4), thus providing the added strength of lignified pegs.

In tobacco and tobacco-tomato grafts (Fig. 6) and in *Coleus* stem grafts (Fig. 7, Stoddard and McCully, 1981) cambial activity begins relatively late (approximately 8 days after grafting in the former plants, 12 days in *Coleus*) and continues for the life of the grafted plant. This activity (which continuously adds lignified elements across the graft) must also increase the tensile strength of the graft union. Cambial activity could thus correlate with the third or levelling off phase of graft development recognized by Moore (1983), although in *Sedum* grafts cambial activity is not mentioned as being present.

In *Coleus* stem grafts older than 10 days there is also the development of lignified secondary walls in the pith parenchyma cells adjacent to and within several cell lengths of the graft line (McIntyre, McCully and Stoddard, in preparation). This activity would also be expected to add strength to the graft union.

Our initial studies of all these grafts and of the pea root grafts have been with hand sections of fresh material, and have emphasized to us the heterogeneity of events along the graft line. For example, the sequence of structural changes described above for the stem grafts of tomato and tobacco occurs mainly in those regions close to the original vascular strands of stock and scion. The pith parenchyma region may also form an apparently strong union, although there is little or no callus proliferation (Fig. 2) or vascular differentiation in this part of the graft. Grafts of monocotyledons resemble pith parenchyma grafts in having the same relatively simple union (at least early in their development) (Muzik and La Rue, 1954; Muzik, 1958; Fig. 8). Furthermore, in the *in vitro* grafts between *Coleus* internodes (Stoddard and McCully, 1981), neither cellular division nor vascular differentiation is involved in graft development, yet the partners adhere tenaciously. Of course, post-genital fusions (see Walker, this volume) provide the best example of cohesion in the absence of cellular division, vascular differentiation or even tissue wounding.

It would be possible to follow the development of graft strength in the simple grafting situations, particularly in the *Coleus* internode system of Stoddard and McCully (1981). Parkinson and Yeoman (1982) have shown that stem internodes grafted *in vitro* do develop cohesive strength in a two phase pattern, but this grafting was in the presence of added plant hormones and involved both callus production and vascular redifferentiation. If a complex pattern of development of tensile strength was also a feature of the simple internode grafts, then considerations of stock and scion cohesion would have to focus on the development of the necrotic layer, the secretions into it, and its structural and chemical transformation. These events must effect graft cohesion, and may play a more important role than has been suspected in later stages of graft development. Furthermore, these events may possibly be involved in graft incompatibility (see below).

Cut plant tissues produce exudates that may act as glue. Moore and Walker (1981b) have demonstrated that a scion will adhere to a wooden applicator stick. The presence of the beaded structures (Fig. 1) on the surface of callus cells invading the graft (see also Fig. d, plate 1 of Yeoman and Brown, 1976) suggests, but in no way proves, a localized secretion of a polyphenolic substance. In this regard, Moore (1982) and Moore and Walker (1981a) have observed a marked increase in the number of dictyosomes in surface cells early in graft development, and have suggested that these organelles are involved in secretion of wall material that acts as a glue. This may well be the case, but such speculation should await some clarification of the nature and the source of the glue, for in such rapidly expanding cells dictyosome activity could be involved only in synthesis of new wall material.

The best histochemical characterization of material released from a cut plant surface is that of Romberger and Tabor (1975), who found that excised shoot apices of *Picea abies* extruded precursors for water insoluble substances that subsequently polymerized as far as 200 μm from the nearest cellular membrane. Periodic acid-Schiff's positive and pectin-like polysaccharides were present in the externally synthesized polymers as well as compounds giving positive reactions for lignin and protein. It is particularly interesting that the external polymers could be detected as early as 3 hours after wounding, and that exudation continued for 9 to 15 days. Similar continuing polymerization of secretions by cells at or close to the graft line could initially fuse cells together and increase the tensile strength of the graft union with time. Toluidine blue staining reactions certainly indicate the presence of polyphenolic compounds in the necrotic material at early stages of successful grafts. These compounds may well effect fusion by linking secreted polysaccharides and existing collapsed walls. Deloire and Hébant (1982) have demonstrated positive histochemical reactions for both lignin and peroxidase activity in the fusion line in the early stages of successful graft development in *Capsicum* and *Lycopersicum*. However, older, successful grafts are characterized by the disappearance of these histochemical reactions. Similarily, in successful grafts of tomato and tobacco stems and in pea root grafts (Stoddard and McCully, 1979), toluidine blue staining for polyphenols at the fusion line is replaced during development by a strong positive staining for pectins. Clearly, the chemistry of graft fusion and its changes during development needs investigation.

Bridging of the graft by a vascular cambium and the subsequent production of secondary vascular tissues—The production of secondary vascular tissues in continuity with similar tissues of the stock and scion and the continued matching of cambial activity of stock and scion throughout the life of the plant is clearly of considerable importance for the success of woody grafts (see Kramer and Kozlowski, 1979). However, the differentiation of cambial initials from callus cells at the graft union and the initiation of cambial activity occur relatively late in graft development. Furthermore, continuity of secondary tissues never occurs in grafts between monocotyledons. Thus, cambial activity is relatively unimportant in the grafting process *per se*, but of utmost importance in the maintenance of an integrated growth pattern between graft partners in which secondary growth is a normal developmental feature.

The orientation of the new cambium relative to that of the original cambia is of obvious importance in the reestablishment of successfully integrated growth. The course of cambial differentiation across grafts with different orientations has been extensively studied by Warren Wilson and Warren Wilson (for example, see their 1981 paper) and Thair and Steeves (1976).

IS THERE SYMPLASTIC CONTINUITY ACROSS A GRAFT UNION?

While experimental studies of physiological and morphogenetic phenomena associated with grafting (Zeevaart, 1976; Sachs, 1981) and of the transmission of viruses across grafts (Fridlund, 1979) have assumed the presence of plasmodesmatal connections, structural evidence for such connections is based entirely upon a few old observations made with optical microscopy. Interestingly, there is considerable discussion in the literature concerning the accuracy of these observations (see Carr, 1976; Jones, 1976).

Recently, McIntyre, McCully, and Stoddard (in preparation) have investigated the fine structure of the graft line between pith parenchyma regions of *Coleus* stem autografts. The small amount of callus production in these locations makes it possible to distinguish clearly the graft line. Parenchyma cells of each partner adjoining the line do develop sizeable pits in their thickened secondary walls. However, serial sections through these pits show no bridging plasmodesmata. In this regard, stable periclinal chimeras (Neilson-Jones, 1969), which could be considered the ultimate in a successful graft, would certainly be expected to show normal plasmodesmatal connections. However, Burgess (1972) found that in *Cytisus adami*, walls linking cells of the two partners are pitted, but plasmodesmata transverse only half of the cell wall, and have no direct continuity across it.

Sieve tube strands which bridge a graft union must provide localized regions of plasma membrane continuity. However, present evidence suggests that other symplastic connections may not necessarily be present. This important point needs further investigation.

THE STRUCTURE OF INCOMPATIBLE GRAFTS

Most of the literature dealing with the structure of unsuccessful grafts describes events which occur late in graft formation between woody plants (see discussion and references in Kramer and Kozlowski, 1979; Moore, 1981). If there are specific structural features resulting from inherent tissue incompatibility, these must be looked for at early stages, particularly in grafts of herbaceous plants.

In the grafts which have been studied, the earliest stages of graft development, cell enlargement, callus formation and at least the first phase of the development of cohesive strength are common to both compatible and incompatible grafts (Lauchaud, 1975; Moore and Walker, 1981a; Moore, 1983; Moore, 1981). Deloire and Hébant (1982) have also shown that lignification of the necrotic zone proceeds for the first few days similarly in both types of grafts. Fridlund (1979) has suggested on the basis of viral transfer that symplastic continuity must be established early in the development of both types of grafts (but see discussion above). Later events of successful grafts seem to occur variably in unsuccessful situations. In some cases wound vascular elements may not form, or vascular cambium activity is not initiated (e.g., Deloire and Hébant, 1982). In other situations vascular bridges do form, but sieve tubes subsequently collapse or become lignified (de Stigter, 1959; Schmid and Feucht, 1981). Nothing is known of the factors involved either in the failure of vascular differentiation or the subsequent deterioration of sieve tubes. The very interesting experiments of de Stigter (1959, 1971) suggest that the situation is very complex, since the presence of some stock tissue grafted to what would otherwise be an incompatible scion prevents the expected collapse of the sieve tubes.

Moore and Walker (1981b, 1983) have reported that graft incompatibility between *Sedum* and *Solanum* is characterized by death of *Sedum*, but not *Solanum*, cells in grafts between intact stems as well as callus masses. Ball (1969) earlier showed that cell death occurred at the interface of callus masses of incompatible tissues.

It is difficult to assess the structural appearance or the histochemistry (Moore and Walker, 1981b) done on the dying cells of *Sedum* because of the large amounts of phenolic compounds present. However, some cells do appear to produce suberized wall lamellae. Suberized cells also characterize many unsuccessful grafts (Copes, 1969; Deloire and Hébant, 1982), but these are generally within phellogen and thus do not seem easily relatable to the cellular death observed in *Sedum* cells by Moore and Walker (1981b).

Deloire and Hébant (1982) found that increasing lignification of the graft line characterized the incompatible grafts they studied, whereas lignification ceased early in the development of successful grafts. This observation, together with the observations of transformation in the staining reactions of the graft line (see discussion above) suggests that failure to modify the chemistry of the initial graft cohesion mechanism may ultimately result in graft failure.

ARE THERE STRUCTURAL EVENTS UNIQUE TO GRAFTING?

Most of the structural events of grafting do not appear to be unique to grafting. For example, cellular elongation and callus production at the wounded surface are common plant responses (La Rue, 1937). Tyloses occur frequently following wounding (see discussion in Kramer and Kozlowski, 1979), and are always found during the development of abscission zones (Addicott, 1982).

Necrotic zones with staining reactions indicating the secretion of lignin or lignin-like compounds characterize all wounded plant surfaces (Vance et al., 1980; Pearse and Rutherford, 1981; Fleuriet and Deloire, 1982). Wound vascular elements develop whenever vascular strands are severed (Sachs, 1981), and a vascular cambium will differentiate across wound callus in many situations (see discussion in Thair and Steeves, 1976).

One event which characterizes most wound situations (Lipetz, 1970; Bloch, 1941) is the production of a periderm. This event seems to be absent from successful grafts. This absence may simply be the result of protection of the wounded surfaces from desiccation. However, factors specifically inhibiting this normal response may be involved, as has been shown in the case of some successful pathogen invasions (Hamilton, et al., 1980).

Moore and Walker (1981a) have suggested that cells adjacent to the graft interface in compatible and incompatible grafts undergo a period of 'partial senescence' during graft development. Such a transitory change in cell cytoplasm has not been previously reported and may somehow be unique to grafting. However, because of the fragility and irregularity of the callus cells at a graft line, it is possible that the apparent dilution of the cytoplasm of these cells may be an artifact introduced by damage during specimen preparation. This point should be investigated further.

IS THERE STRUCTURAL EVIDENCE FOR RECOGNITION PHENOMENA IN GRAFTING?

The structural evidence available to date does not suggest a strong possibility of early recognition events at the graft interface based upon contact between cell membranes of the partners. The necrotic layer is always an initial barrier, and there is no evidence for the subsequent development of plasmodesmatal connections. Yeoman, et al. (1978) have suggested that primary recognition between graft partners occurs when plasma membranes come into contact following a temporary phase of localized wall dissolution. The evidence for this event is not compelling (Figs. a & b, plates 1 & 2, Yeoman, et al., 1978). The thickness of the graft line shown in the optical micrographs in each case suggests that the electron micrographs may be from regions not at the graft line. Because of the complexity of cellular interdigitation wherever callus cells fill a graft space, it is almost impossible to interpret resulting electron micrographs. Such a transitory dissolution of cell wall or some transformation in wall structure would seem to be necessary for transmission of early recognition signals. It should be looked for carefully in a situation where the orientation of cells derived from each partner can be determined with certainty.

LITERATURE CITED

Addicott, F. T. (1982). Abscission. University of California Press, Berkeley.
Ball, E. (1969). Histology of mixed callus cultures. Bull. Torrey Bot. Club 96: 52-59.
Bloch, R. (1941). Wound healing in higher plants. Bot. Rev. 7: 110-146.
Burgess, J. (1972). On the occurrence of plasmodesmata-like structures in a non-division wall. Protoplasma 74: 449-458.

Camus, G. (1949). Recherches sur le rôle des bourgeons dans les phenomenes de morphogenese. Revue de Cytol. Biol. Veget. 11: 1-95.

Carr, D. J. (1976). Plasmodesmata in growth and development. In Intercellular Communication in Plants: Studies on Plasmodesmata. Ed. by B. E. S. Gunning and A. W. Robards. Springer Verlag, Berlin, Heidelberg, N. York.

Copes, D. (1969). Graft union formation in the douglas-fir. Amer. J. Bot. 56: 285-298.

Copes, D. L. (1970). Initiation and development of graft incompatibility systems. Silvac Genet. 19: 101-107.

Crafts, A. S. (1934). Phloem anatomy in two species of Nicotiana, with notes on the interspecific graft union. Bot. Gaz. 95: 592-608.

Daniel, L. (1927). Études Sur La Greffe. Vol. 1. Imprimerie Oberthur. Paris.

Deloire, A. and C. Hébant. (1982). Peroxidase activity and lignification at the interface between stock and scion of compatible and incompatible grafts of Capsicum and Lycopersicum. Annals of Bot. 49: 887-891.

De Stigter, H. C. M. (1959). L'incompatibilite partielle dans le greffage moyen d'investigation physiologique. Bull. Soc. Sci. Bretagne 34: 104-108.

De Stigter, H. C. M. (1971). Some aspects of the physiological function of the graft muskmelon/Cucurbita ficifolia. 4. Observations on a Cucurbita isograft "model," and effects of localized stem cooling. Zeit. Pflanzenphysiol. 65: 296-308.

Fleuriet, A. and D. Deloire. (1982). Aspects histochimiques et biochimiques de la cicatristation des fruits de Tomate blesses. Zeit. Pflanzenphysiol. 107: 259-268.

Fridlund, P. R. (1979). Intrageneric graft compatability in relation to graft transmission of Prunus virus. Lucr. Stiint Inst. Agron 'Nicolae Balcescu,' Agron 23: 5-10.

Hamilton, S., E. McGee, M. C. Jarvis and H. J. Duncan. (1980). Effects of Phytophthora infestans on wound healing of potato tuber discs and on their infection with Phoma exiqua. Physiol. Pl. Path. 17: 303-307.

Hayward, H. E. and F. W. Went. (1939). Transplantation experiments with peas. II. Bot. Gaz. 100: 788-801.

Jones, M. G. K. (1976). The origin and development of plasmodesmata. In Intercellular Communication in Plants: Studies on Plasmodesmata. Ed. by B. E. S. Gunning and A. W. Robards. Springer Verlag, Berlin, Heidelberg, N. York. Kramer, P. J. and T. T. Kozlowski. 1979. Physiology of Woody Plants. Academic Press, New York, San Francisco, London.

La Rue, C. D. (1937). Cell outgrowths from wounded surfaces of plants in damp atmospheres. Mich. Acad. Sci. Papers 22: 123-139.

Lauchaud, S. (1975). Incompatibilite des greffes et viellissement chez les vegetaux. 2. L'incompatibilite des greffes se ses rapports avec le vieillissement. Ann. Biol. 14: 98-128.

Lindsey, D. W., M. M. Yeoman and R. Brown. (1974). An analysis of the development of the graft union in Lycopersicon esculentum. Ann. Bot. 38: 639-646.

Lipetz, J. (1970). Wound healing in higher plants. Int. Rev. Cytol. 27: 1-28.

Moore, R. 1981. Graft compatibility and incompatibility. Devel. Comp. Immun. 5: 377-389.

Moore, R. (1982). Studies of vegetative compatibility-incompatibility in higher plants. 5. A morphometric analysis of the development of a compatible and an incompatible graft. Can. J. Bot. 60: 2780-2787.

Moore, R. 1983. Studies of vegetative compatibility-incompatibility in higher plants. 4. The development of tensile strength in a compatible and an incompatible graft. Amer. J. Bot. 70: 226-231.

Moore, R. and D. B. Walker. (1981a). Studies of vegetative compatibility-incompatibility in higher plants. I. A structural study of a compatible autograft in *Sedum telephoides* (Crassulaceae). Amer. J. Bot. 68: 820-830.

Moore, R. and D. B. Walker. (1981b). Studies of vegetative compatibility-incompatibility in higher plants. II. A structural study of an incompatible heterograft between *Sedum telephoides* (Crassulaceae) and *Solanum pennellii* (Solanaceae). Amer. J. Bot. 68: 831-842.

Moore, R. and D. B. Walker. (1981c). Studies of vegetative compatibility-incompatibility in higher plants. 3. The involvement of acid phosphatase in the lethal cellular senescence associated with an incompatible heterograft. Protoplasma 109: 317-334.

Moore, R. and D. B. Walker. (1983). Studies of vegetative compatibility-incompatibility in higher plants. 5. Grafting of *Sedum* and *Solanum* callus tissue *in vitro*. Protoplasma, in press.

Muzik, T. J. 1958. Role of parenchyma cells in graft union in vanilla orchid. Science 127: 82.

Muzik, T. J. and C. D. La Rue. (1952). The grafting of large monocotyledonous plants. Science 116: 589-591.

Muzik, T. J. and C. D. La Rue. (1954). Further studies on the grafting of monocotyledonous plants. Amer. J. Bot. 41: 448-455.

Nielson-Jones, W. (1969). Plant Chimeras. Methuen and Co. London.

Parkinson, M. and M. M. Yeoman. (1982). Graft formation in cultured, explanted internodes. New Phytol. 91: 711-719.

Pearce, R. B. and J. Rutherford. (1981). A wound-associated suberized barrier to the spread of decay in the sapwood of oach (*Quercus robur* L.) Physiol. Pl. Path. 19: 359-369.

Robbertse, J. R., and M. E. McCully. (1979). Regeneration of vascular tissue in wounded pea roots. Planta 145: 167-173.

Roberts, J. R. and R. Brown (1961). The development of the graft union. J. Exp. Bot. 12: 294-302.

Romberger, J. A. and C. A. Tabor. (1975). The *Picea abies* shoot apical meristem in culture. 2. Deposition of polysaccharides and lignin-like substances beneath cultures. Amer. J. Bot. 62: 610-617.

Sachs, T. (1981). The control of patterned differentiation of vascular tissues. *In* Advances in Botanical Research, Vol. 9. Ed. by H. W. Woolhouse. Academic Press, New York. pp. 151-262.

Sass, J. E. (1932). Formation of callus knots on apple grafts as related to the histology of the graft union. Bot. Gaz. 94: 364-380.

Schmid, P. P. S. and W. Feucht. (1981). Differentiation of sieve tubes in compatible and incompatible *Prunus* graftings. Sci. Hortic. (Amst.) 15: 349-354.

Simon, S. (1930) . Transplantation Versuche zwischen *Solanum melongena* und *Iresine lindeni*. Jahrb. Wiss. Bot. 72: 137-160.

Stoddard, F. L. 1981. Comparative Effects of Stock and Scion Organs and of Graft Partners on the Developmental Anatomy of the Graft Union in *Coleus*. M.Sc. thesis. University of Ottawa.

Stoddard, F. L. and M. E. McCully. (1979). Histology of the development of the graft union in pea roots. Can. J. Bot. 57: 1486-1501.

Stoddard, F. L. and M. E. McCully. (1980). Effects of stock and scion organs on the formation of the graft union in *Coleus*: A histological study. Bot. Gaz. 141: 401-412.

Thair, B. W. and T. A. Steeves. (1976). Response of the vascular cambium to reorientation in patch grafts. Can. J. Bot. 54: 361-373.

Vance, C. P. (1980). Lignification as a mechanism of disease resistance. Ann.

Rev. Phytopath. 18: 259-288.

Yeoman, M. M. and R. Brown (1976). Implications of the formation of the graft union for organization in the intact plant. Ann. Bot. 40: 1265-1276.

Yeoman, M. M., D. C. Kilpatrick, M. B. Miedzybrodzka and A. R. Gould. (1978). Cellular interactions during graft formation in plants, a recognition phenomenon? Symps. Soc. Exp. Biol. 32: 139-159.

Warren Wilson, J. and P. M. Warren Wilson. (1981). The position of cambia regenerating in grafts between stems and abnormally oriented petioles. Ann. Bot. 47: 473-484.

Zeevaart, J. A. D. (1976). Physiology of flower formation. Ann. Rev. Plant Physiol. 27: 321-348.

VI

Physiological Aspects of Graft Formation

Randy Moore

Department of Biology, Baylor University
Waco, TX 76798

INTRODUCTION

When unrelated plants are grafted together, the result is usually an unsuccessful graft. Conversely, when closely related plants are grafted, they typically unite readily and begin to grow as a single organism. Such congeniality between different plants has not gone unnoticed by botanists, who have exploited tissue compatibility to "engineer" plants beneficial to humankind. For example, *Vitis vinifera* (European grape) is typically susceptible to grape phylloxera (Mahlsted and Haber, 1957). However, when *Vitis lambrusca* (American grape) is used as a rootstock for *V. vinifera*, *V. vinifera* becomes resistant to grape phylloxera, and there is a resulting increase in yield (Mahlsted and Haber, 1957; Hartmann and Kester, 1975). Plant grafting is also used to obtain special forms of growth, to perpetuate and speed production of certain clones of plants, to improve fruit quality and winter-hardiness, and to facilitate the adaptation of plants to different environments (Mahlsted and Haber, 1957; Hartmann and Kester, 1975).

The recent renewal of interest in plant grafting has resulted in an increased understanding of the processes associated with graft formation (see reviews by Moore, 1981a, 1981b). Most studies have concentrated on structural aspects of graft development (Copes, 1969, 1980; Shimomura and Fuzihara, 1976; Stoddard and McCully, 1979; Moore, 1982, In Press a; Moore and Walker, 1981a, 1981b). However, other investigations have addressed questions relating to physiological aspects of graft formation (Shimomura and Fuzihara, 1977, 1978; Copes, 1978; Yeoman, et al., 1978; Stoddard and McCully, 1980; Moore and Walker, 1981c, In Press; Moore, In Press b). For example, what factors are involved in re-establishing vascular continuity between graft partners? What roles do plant growth regulators play in graft formation? What mechanisms determine graft compatibility-incompatibility?

GRAFT COMPATIBILITY

Developmental Aspects

Compatible grafts are characterized by a strikingly similar pattern of development (Fig. 1). In every system, the increase in tensile strength of

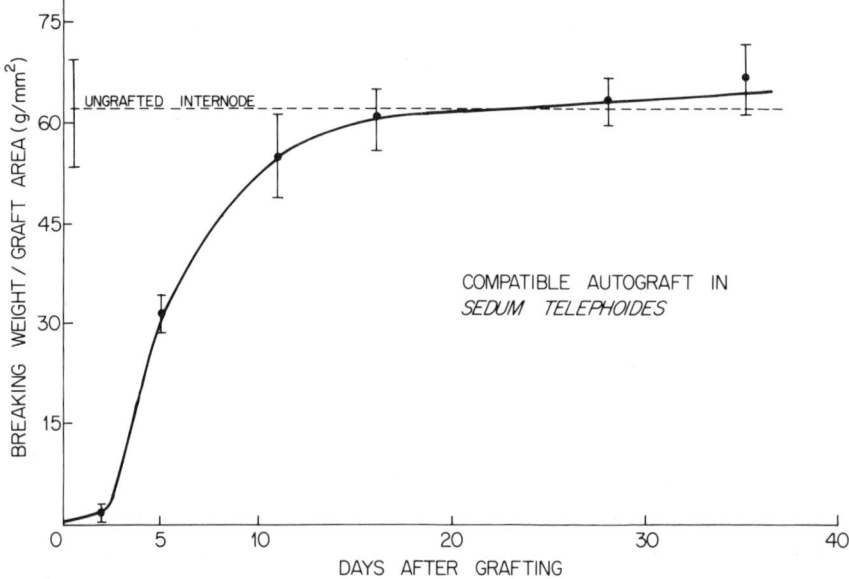

FIGURE 1. *The development of tensile strength in compatible autografts in* Sedum telephoides. *It appears that all compatible autografts have a similar pattern of development of tensile strength. (From Moore, In Press b)*

the graft union is positively correlated with three developmental events: (1) cohesion of the stock and scion, (2) proliferation of callus tissue, and (3) redifferentiation of vascular tissue across the graft interface (Moore, 1982, In Press b).

Cohesion of the Stock and Scion

The initial cohesion of the stock and scion is the result of a cellular wound response induced in grafting cells by the graft incision. The dictyosome-mediated secretion (and subsequent polymerization) of cell wall precursors into the graft interface appears to be the basis for cohesion of the graft partners (Fig. 2; Moore and Walker, 1981a, 1981b;

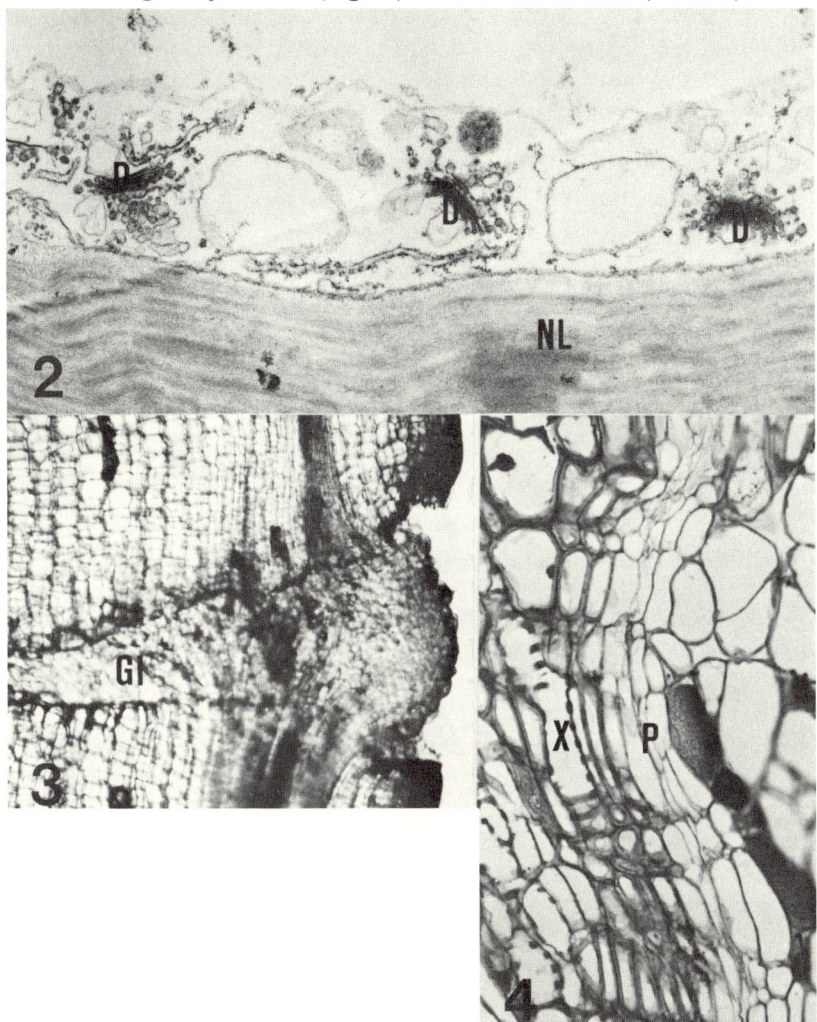

FIGURE 2. Accumulation of dictyosomes (D) along the cell walls adjacent to the necrotic layer (NL) at 6 hrs after grafting in the compatible Sedum autograft. X21,300. (From Moore and Walker, 1981a)

FIGURE 3. Callus cells at the graft interface (GI) of the compatible Kalanchoe autograft at 8 days after grafting. X40. (From Moore, 1982)

FIGURE 4. Vascular continuity between the stock and scion of the compatible Sedum autograft at 14 days after grafting. P=phloem; X=xylem. X400. (From Moore and Walker, 1981a)

Moore, 1982). The initial cohesion of the stock and scion does not involve mutual cellular recognition (Moore and Walker, 1981a, 1981b) and is independent of other events in graft formation (Stoddard and McCully, 1980). Also, since the stock and scion of incompatible grafts typically adhere during the early stages of graft formation (Yeoman, et al., 1978; Moore and Walker, 1981b), the initial cohesion of graft partners is not directly related to graft compatibility-incompatibility.

Callus Proliferation

Callus proliferation (Fig. 3) is believed to be essential for the formation of a successful graft (Hartmann and Kester, 1975). Like the initial cohesion of the stock and scion, callus proliferation is independent of other events in graft formation (Stoddard and McCully, 1980). Since callus proliferation occurs in compatible and incompatible grafts (Moore and Walker, 1981a, 1981b) as well as in nongrafted systems (Barckhausen, 1978), callus formation (1) is not directly related to graft compatibility-incompatibility, and (2) does not involve mutual cellular recognition.

Redifferentiation of Vascular Tissue

Redifferentiation of vascular tissue across the graft interface (Fig. 4) is one of the last major events to occur during the formation of a compatible graft (Stoddard and McCully, 1979; Moore and Walker, 1981a; Moore, 1982). Vascular redifferentiation is independent of other events in graft formation (Stoddard and McCully, 1980). Also, cellular recognition is probably involved in at least one aspect of vascular redifferentiation during graft formation (see below).

It has been suggested that the redifferentiation of vascular tissue between the stock and scion is the critical structural event in the formation of a successful graft (Yeoman, et al., 1978). However, there are reports of survival of scions in the absence of vascular continuity (Herrero, 1951; Muzik, 1958; Moore, unpublished data). Therefore, redifferentiation of vascular tissue is not necessarily related to a successful graft, much less to the compatibility-incompatibility of the tissues involved (Moore and Walker, 1981a).

The Roles of Plant Growth Regulators in Graft Development

Our knowledge of the roles of plant growth regulators in grafting is limited to their involvement in the formation of compatible unions. Several investigators have reported that graft formation is promoted by the application of auxin (Kruyt, 1847; Muller-Stoll, 1938; Shimomura and Fuzihara, 1977, 1978; Parkinson and Yeoman, 1982). For example, Muller-Stoll applied auxins to the grafting surfaces of grape vines and demonstrated that a 0.05% solution of indoleacetic acid (IAA) is most effective at promoting graft development. Kruyt (1847) reported similar results with applications of naphthalene acetic acid (NAA) to the grafting surfaces of ornamental shrubs. In these systems, auxin promotes graft formation by stimulating the production of callus tissue at the graft interface. Indeed, an auxin gradient is necessary for callus proliferation (Jacobs, 1979).

Application of auxin to scion apices promotes the redifferentiation of vascular tissue in cactus grafts (Shimomura and Fuzihara, 1977). Auxin-promoted redifferentiation of vascular tissue is inhibited by triiodobenzoic acid (TIBA), an inhibitor of basipetal transport of auxin. When auxin is applied to a stock instead of a scion, a cambium forms in the tissue between the point of application and the severed vascular bundle. A similar effect has also been reported in other systems, in which the insertion of a bud into callus tissue induced the differentiation of vascular tissue (Camus, 1949). Since an auxin gradient is necessary for regeneration of vascular tissue (Jacobs, 1952; Sachs, 1968), vascular differentiation across a graft interface may be controlled by endogenous auxin of the scion (Shimomura and Fuzihara, 1977). Auxin contributed by the scion must be supplied continuously (Stoddard and McCully, 1980), since auxin destroying enzymes accumulate at cut surfaces within 6 hours (Iversen and Aasheim, 1970). Also, auxin has no effect on graft cohesion in unsuccessful heterografts between cacti (Shimomura and Fuzihara, 1977), consistent with the suggestion that cohesion of the stock and scion is due to a wound response (and not directly related to graft compatibility-incompatibility).

The observations of Shimomura and Fuzihara (1977) outlined above are supported by the fact that the amount of graft-bridging xylem is strongly influenced by the presence of leaves and branches on the scion, but not at all by their presence on the stock (Stoddard and McCully, 1980). Estimates of the contributions of xylem-inducing stimuli by plant organs in grafts between *Coleus* stems are given in Table 1. Organs

---

TABLE 1. Contributions of xylem-inducing stimuli by plant organs in *Coleus* autografts. (From Stoddard and McCully, 1980).

Acropetal flow from stock to scion	8%
Basipetal flow from immature stem tissue	10%
Shoot tips and immature leaves on scion	10%
Scion leaves that are at least half mature	36%
Synergistic effects of scion organs (?)	36%

---

which produce the most auxin (i.e., leaves and shoot tips, Jacobs, 1952; Scott and Briggs, 1960; Thimann, Sachs, and Mathur, 1971) are the most effective organs at inducing callus proliferation and vascular redifferentiation in plant grafts (Stoddard and McCully, 1980).

Parkinson and Yeoman (1982) have reported that application of IAA is an absolute requirement for the formation of a successful graft in cultured internodes of *Lycopersicon, Datura,* and *Nicandra.* The presence of kinetin in the culture medium increases the amount of vascular tissue bridging the graft union, but only in the presence of auxin. Gibberellic acid (GA) inhibits vascular redifferentiation in this *in vitro* grafting system (Parkinson and Yeoman, 1982). There is disagreement in the literature on the roles of cytokinins and GA in vascular differentiation (Fosket and Roberts, 1964; Roberts and Fosket, 1966; Waisel, Noah, and Fahn, 1966; Earle, 1968; Harrison and Klein, 1979).

Since there is very little callus proliferation in the stock until cohesion, hormones originating in the stock and wound hormones (Zimmerman and Coudron, 1979) can be eliminated as prompt eliciters of cellular division during the early stages of graft formation (Stoddard and McCully, 1980). The delayed cellular division and vascular redifferentiation in the stock appear to be due to hormones originating in the scion which move across the graft interface (Stoddard and McCully, 1980). The wound hormone traumatin may be involved in the induction of cohesion between stock and scion in *Coleus* grafts (Stoddard and McCully, 1980).

GRAFT INCOMPATIBILITY

An "incompatible" graft is not synonymous with an "unsuccessful" graft. For example, the adjacent tissues of a severed (and regrafted) internode are certainly compatible. However, such autografts are occasionally unsuccessful. The failure of graft development in these instances is typically due to poor grafting technique rather than tissue incompatibility. Also, certain graft combinations (e.g., pear/quince) exhibit "degrees" of incompatibility (Gur, Samish, and Lifschitz, 1968). This observation, combined with the fact that there are numerous causes for an unsuccessful graft (Table 2; see review by Moore, 1981b) led Moore and Walker (1981b) to restrict their definition of graft incompatibility to those instances in which mutual physiological influences (or lack thereof) between tissues of the stock and scion ultimately result in an unsuccessful graft.

While it appears that the general sequence of events mentioned above (i.e., cohesion of stock and scion, callus proliferation, and redifferentiation of vascular tissue) occurs in response to grafting compatible tissues, there are several different responses that can result in an unsuccessful graft (Table 2). Incompatibility responses have been categorized as (1)

TABLE 2. Structural features correlated with graft incompatibility.

TYPE OF RESPONSE	REFERENCE
Absence of cellular redifferentiation and subsequent formation of callus tissue	Hartmann and Kester, 1975
Lack of vascular redifferentiation	Muzik, 1958
Regeneration of phloem	Mosse, 1962
Lack of phloem redifferentiation	McClintock, 1948
Degeneration or redifferentiation of the vascular cambium	Herrero, 1951 Bradford and Sitton, 1929
Inadequate fusion of cambial initials	Eames and Cox, 1945
Cellular necrosis at the graft interface	Buck, 1954 Copes, 1980 Fletcher, 1964 Gur, Samish, and Lifschitz, 1968 Moore and Walker, 1981b

translocated incompatibility, (2) localized incompatibility, and (3) delayed incompatibility (Hartmann and Kester, 1975).

Translocated incompatibility is the type of incompatibility response that is not overcome by the insertion of a mutually compatible interstock (e.g., almond/plum). One possible explanation for translocated incompatibility is that a toxin from one partner moves through the interstock and adversely affects the other graft partner.

Localized incompatibility is not overcome by the insertion of a mutually compatible interstock (e.g., certain varieties of pear/quince). In these systems, the interstock apparently acts as a trap for substances originating in one partner which elicit incompatibility in the other partner.

Delayed incompatibility is the type of incompatibility response that develops only after a period of up to 20 years after grafting (e.g., *Juglans regia/J. hindsii*). Possible explanations for delayed incompatibility include viral infections (Hartmann and Kester, 1975) and the production of a toxic secondary metabolite in response to aging.

Cellular Necrosis and Graft Incompatibility

The incompatibility response that occurs in each of the cases mentioned above (i.e., translocated, localized, and delayed incompatibility) is cellular necrosis. The only grafting systems in which mechanisms for graft-induced cellular necrosis have been elucidated are those between (1) pear/quince and (2) peach/almond.

Quince (*Cydonia oblonga*) rootstocks contain prunasin, a cyanogenic glycoside (Gur, Samish, and Lifschitz, 1968; Gur, Zamet, and Arad, 1978). When grafted to pear (*Pyrus communis*), which lacks prunasin, the prunasin from quince ascends into the pear scion, where it is broken down by glycosidase, thus liberating hydrocyanic acid and benzaldehyde at the graft interface. Hydrocyanic acid induces cellular necrosis at the graft interface, thus bringing about graft incompatibility (Fig. 5). Evidence

PEAR / QUINCE INCOMPATIBILITY

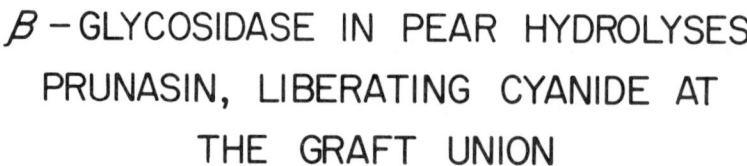

PRUNASIN ASCENDS FROM QUINCE INTO PEAR

β – GLYCOSIDASE IN PEAR HYDROLYSES PRUNASIN, LIBERATING CYANIDE AT THE GRAFT UNION

CYANIDE CAUSES DEVELOPMENTAL ABNORMALITIES AT THE GRAFT INTERFACE

FIGURE 5. *The mechanism for graft incompatibility between pear* (Pyrus communis) *and quince* (Cydonia oblonga) *as proposed by Gur, et al. (1968)*

for this mechanism for graft incompatibility between pear and quince includes the facts that: (1) Movement of prunasin from quince into pear is readily demonstrable after grafting, (2) The quality of the graft union is proportional to glycosidase activity in pear tissue at the graft union, (3) As a result of phloem degeneration, less sugar reaches quince roots, thus leading to further decomposition of prunasin and more cellular necrosis, (4) Some varieties of pear contain a water-soluble inhibitor of glycosidase. This may explain why certain pear varieties graft successfully with quince, while others do not. Also, the increased incompatibility between pear and quince under hot conditions is positively correlated with the fact that the glycosidase responsible for the liberation of hydrocyanic acid (and, thus, cellular necrosis) in pear tissue has a temperature optimum of 35° C (Gur, Samish, and Lifschitz, 1968). Significantly, graft incompatibility between pear and quince has been overcome experimentally by the injection of ferrous polysulphide into the graft interface (Gur, 1972). Ferrous polysulphide inactivates glycosidase, thus preventing the liberation of hydrocyanic acid at the graft interface (Fig. 6).

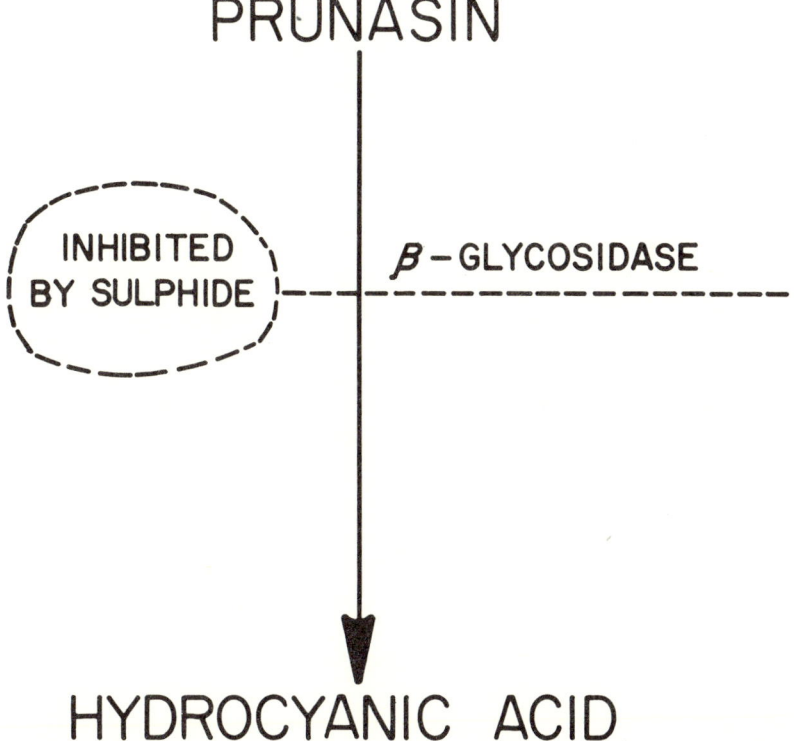

FIGURE 6. *Inactivation of glycosidase by polysulphide. Treatment of pear/quince grafts with ferrous polysulphide has been used to overcome graft incompatibility.*

Incompatibility between peach (*Prunus persica*) and almond (*Prunus amygdalus*) has a biochemical basis similar to that of the pear/quince graft combination. In this system, however, both graft partners contain a cyanogenic glycoside (Gur and Blum, 1973). Since peach contains considerably more of the glycoside than almond, there is a net movement of the glycoside from peach into almond. The resulting increased amounts of cyanogenic glycoside in almond are positively correlated with increased activity of glycosidase in almond (Gur and Blum, 1973). Cyanide liberated at the graft interface induces cellular necrosis, and ultimately brings about graft incompatibility.

Lesser amounts of information are available on mechanisms underlying graft incompatibility in other systems. In the incompatible heterograft between *Sedum telephoides* and *Solanum pennellii*, *Solanum* cells do not exhibit any signs of incompatibility in response to grafting with *Sedum* (Moore and Walker, 1981b). However, *Sedum* cells undergo necrosis in response to grafting with *Solanum*, and thus contribute to a necrotic layer that separates the two graft patners (Fig. 7). Necrosis of *Sedum* cells is similar to the hypersensitive response (Moore and Walker, 1981b), suggesting that cellular necrosis may be a common defense mechanism employed by plants under a variety of conditions. Necrosis of *Sedum* cells is positively correlated with the release of the hydrolytic enzyme acid phosphatase into the cytosol of *Sedum* cells (Figs. 8, 9; Moore and Walker, 1981c). Also, graft incompatibility between *Sedum* and *Solanum* occurs *in vitro* as it does *in vivo* (Moore and Walker, In Press). These results have led Moore and Walker (1981b) to suggest that graft incompatibility between *Sedum* and *Solanum* may be due to toxin(s) that moves from *Solanum* into *Sedum* and there elicits cellular necrosis. The initial site of action of the presumed toxin in the *Sedum/ Solanum* graft combination is the endomembrane system of *Sedum* cells (Moore, In Press a).

The incompatible heterograft between lemon (*Citrus limon*) and *Poncirus trifoliata* is also characterized by extensive cellular necrosis at the graft interface. Bevington (1976) suggested that incompatibility in this system may be due to the fact that some substance produced in lemon moves into and injures tissues of *P. trifoliata*.

Copes (1978) has suggested that 90-100% accuracy is possible in predicting the success of Douglas fir (*Pseudotsuqa manziesii*) grafts by assaying the activity of esterase and peroxidase. Activity of these enzymes increases considerably in unsuccessful grafts as well as in the hypersensitive response (Copes, 1978). Copes' method for predicting the success of a Douglas fir graft is valid only after grafting (Copes, 1978).

Deloire and Hebant (1982) have also reported that peroxidase activity is greater in incompatible grafts between *Capsicum annuum* and *Lycopersicon esculentum* than in corresponding compatible grafts. This in-

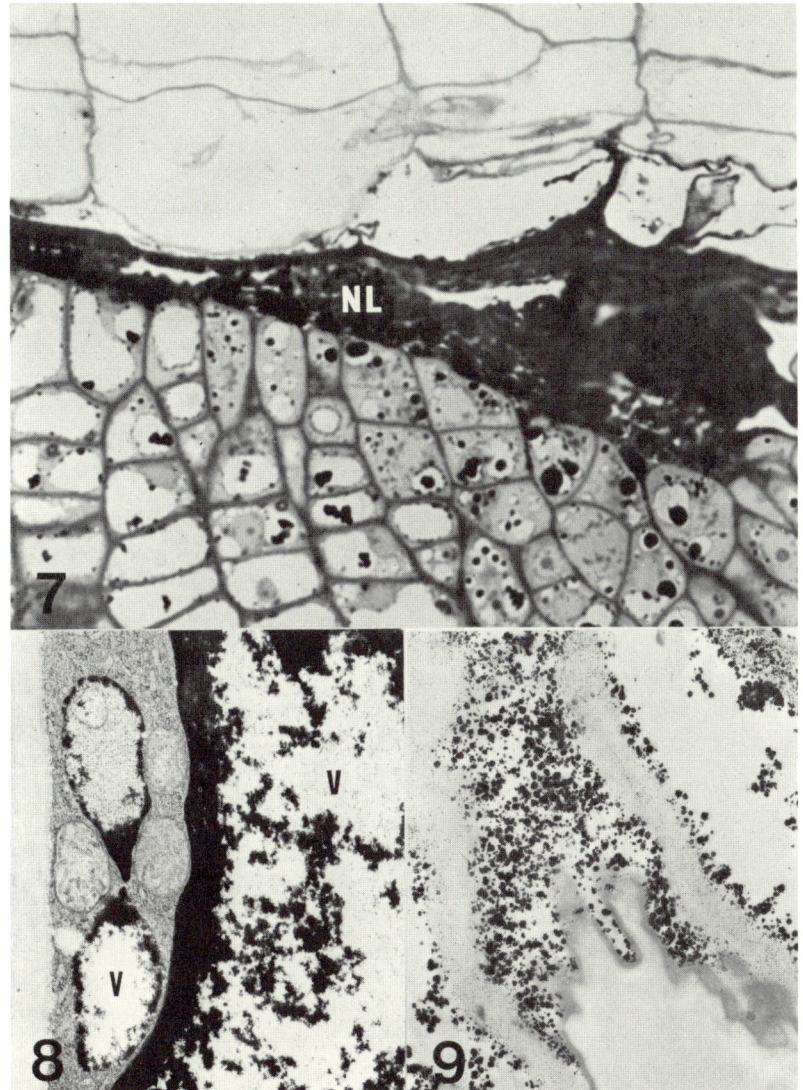

FIGURE 7. The graft interface between Sedum (upper) and Solanum (lower) at 8 days after grafting. Cellular necrosis in Sedum has resulted in the formation of a necrotic layer (NL) separating the graft partners. X850. (From Moore and Walker, 1981b)

FIGURE 8. Acid phosphatase activity along the tonoplast and in the vacuole (V) of Solanum cells near the Sedum/Solanum interface 14 days after grafting. X17,500. (From Moore and Walker, 1981c)

FIGURE 9. Diffuse localization of acid phosphatase in Sedum cells near the Sedum/Solanum interface 14 days after grafting. X31,600. (From Moore and Walker, 1981c)

creased activity of peroxidase may be related to lignification at the graft interface, which is more extensive in incompatible grafts than in compatible ones. Interestingly, Deloire and Hebant (1982) also noted the similarity between graft compatibility-incompatibility and host-parasite interactions in the Solanaceae.

As already mentioned, an unsuccessful graft can result from any of several factors. This suggests that in the graft incompatibility response, partners may reject each other at any of several developmental stages (Stoddard and McCully, 1980). Supporting this claim is the observation that the grafting processes are composed of several superimposed, yet independent, processes (Stoddard and McCully, 1980).

MECHANISMS FOR GRAFT COMPATIBILITY-INCOMPATIBILITY

Yeoman, et al. (1978) have recently suggested that graft compatibility-incompatibility is determined by a mutual cellular recognition system. According to these investigators, "As cells from opposing surfaces touch, the dissolution of the opposing walls is initiated, a hole appears rapidly, and the plasmalemmas come into contact. This sequence occurs whether the parts are compatible or incompatible" (Yeoman, et al., 1978). Protein molecules released from the plasmalemmas then combine to form a "catalytic complex" which initiates a developmental sequence that ultimately determines the success of a graft (Fig. 10).

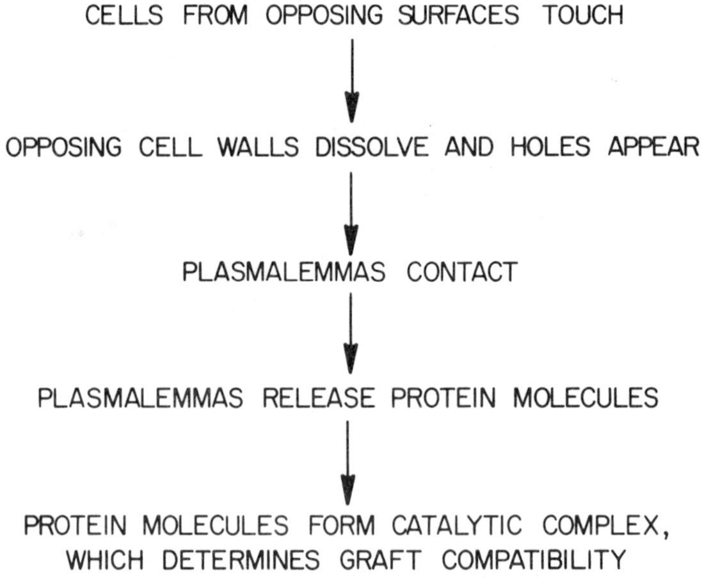

CELLS FROM OPPOSING SURFACES TOUCH

↓

OPPOSING CELL WALLS DISSOLVE AND HOLES APPEAR

↓

PLASMALEMMAS CONTACT

↓

PLASMALEMMAS RELEASE PROTEIN MOLECULES

↓

PROTEIN MOLECULES FORM CATALYTIC COMPLEX,
WHICH DETERMINES GRAFT COMPATIBILITY

FIGURE 10. The mechanism for graft compatibility-incompatibility as proposed by Yeoman, et al. (1978)

There are several problems with this mechanism for graft compatibility-incompatibility. First, the authors provide no micrographs to document the contact of plasmalemmas from opposing surfaces. Indeed, the only data supporting the appearance of a "hole" in the walls of contacting cells at the graft interface is a micrograph of a perforation plate of a xylary element (i.e., a dead cell) (Plate 1C, Yeoman, et al., 1978). Such a dead cell would not have a functional plasmalemma, and hence could not be involved in the cellular recognition process proposed by the authors Secondly, other more comprehensive structural studies of graft development (Stoddard and McCully, 1979; Moore and Walker, 1981a, 1981b; Moore, 1982) have not reported cell wall dissolution to occur in response to grafting. On the contrary, some of these studies have reported deposition of cell wall materials to occur in response to grafting (Moore and Walker, 1981a; Moore, 1982), a common wound response in plants (see review by Aist, 1976) that would have an effect opposite that of the proposed dissolution of cell wall. Thirdly, it is difficult to explain the differential responses of reciprocal grafts with a molecular-level type of recognition system. In these grafts, the A/B (i.e., scion/stock) combination is compatible, but the B/A combination is unsuccessful (Hartmann and Kester, 1975). It is difficult to envision how simply reversing the orientation of a graft combination would significantly alter any molecular-level interactions between contacting surfaces. A simpler explanation for the differential responses of reciprocal grafts is that certain factor(s) produced in one partner only (in this case B) are translocated downward from the B scion and bring about incompatibility in A.

An alternate mechanism for graft compatibility has been proposed by Moore and Walker (1981a). These authors do not believe that mutual cellular recognition (i.e., involving lectins, glycoprotein receptors, etc.) is involved in either the initial cohesion of graft partners or the proliferation of callus tissue at the graft interface (see discussion above). Indeed, both of these aspects of graft development appear to be the result of the wounding necessary to make the graft. Redifferentiation of vascular tissue also occurs in response to wounding when vascular strands are severed (see discussions in Hardham and McCully, 1982a, 1982b). During the course of wound-induced vascular redifferentiation, the differentiation of superimposed cellular structures such as sieve plates and sieve pores, plasmodesmata, and pits must involve some form of cellular recognition and/or communication. Furthermore, the coordinated differentiation of vascular strands may also require some form of cellular recognition. However, the involvement of a separate morphogenetic determinant (e.g., a hormonal gradient) cannot be ruled out (Moore and Walker, 1981a). Indeed, the position and orientation of the cambium in grafts between stems and petioles is believed to be controlled by the ratio of two diffusible morphogens, possibly sucrose and auxin (Warren

Wilson, 1978; Warren Wilson and Warren Wilson, 1981).

The precise involvement of cellular recognition in incompatible grafting is more difficult to determine, since there are several causes of graft incompatibility. While cellular recognition may be involved in determining graft incompatibility between some combinations of plants, it is not necessary to assume that a recognition phenomena is operative in all incompatible grafts. For example, graft incompatibility between pear and quince (see above) can be explained without invoking cellular recognition.

DIRECTIONS FOR FUTURE RESEARCH

In spite of the economic importance of plant grafting, our knowledge of the mechanisms directing graft compatibility-incompatibility is limited. Important questions that await research include: Are the morphogenetic determinants that direct the development of a compatible graft absent in an unsuccessful graft, or are the grafting cells somehow rendered unresponsive to their presence? What are the mechanisms responsible for the differential responses of certain clones within a species to grafting (Copes, 1970)? For reciprocal grafts (Heppner and McCallum, 1927; McClintock, 1948)? For age-related incompatibility (Herrero and Tabuenca, 1969)? Finally, can graft incompatibility be overcome by experimental manipulation?

There is a clear need for more research aimed at elucidating biochemical mechanisms underlying graft incompatibility. Toxic substances (e.g., prunasin) undoubtedly play important roles in eliciting graft incompatibility between certain plants. However, it is unlikely that there is a universal mechanism for graft incompatibility, since there are numerous causes for an unsuccessful graft.

Although plant grafts have been used to study transmission of the flowering stimulus (Lang, Chailakhyan, and Frovola, 1977) and viruses (Hartmann and Kester, 1975) between plants, other observations suggest that grafts may be a powerful tool for studying several other aspects of plant growth and development. For example, when the Baldwin variety of apple is grafted as a scion on French crab apple, both partners are sensitive to severe winter cold. When the hardy Oldenberg variety is grafted on top of the Baldwin variety (to yield an Oldenberg/Baldwin/French crab apple tree), the entire plant is resistant to cold temperatures (Mahlsted and Haber, 1957). Such a system could be important for determining how to induce winter-hardiness in other plants. Other areas in which plant grafts may be successfully used as a research tool include root and shoot development (Hartmann and Kester, 1975), leaf drop, and plant senescence (Mahlsted and Haber, 1957).

LITERATURE CITED

Aist, J. R. 1976. Papillae and related wound plugs. Ann. Rev. Phytopathol. 14: 145-163.

Barckhausen, R. 1978. Ultrastructural changes in wounded plant storage tissue cells. In G. Kahl (ed.), Biochemistry of wounded plant tissues, p. 1-42. Walter de Gruyter and Co., Berlin.

Bradford, F. C. and B. G. Sitton. 1929. Defective graft unions in the apple and the pear. Tech. Bull. Agr. Expt. Sta. Michigan State Coll. 99: 106.

Buck, G. J. 1954. The histology of the bud graft union in roses. Iowa State College J. Sci. 28: 587-602.

Camus, G. 1949. Recherches sur le role des bourgeons dan les phenomenes de morphogenese. Rev. Cytol. Biol. Vegetale 11: 1-199.

Copes, D. 1969. Graft union formation in Douglas fir. Amer. J. Bot. 56: 285-289.

_____. 1970. Initiation and development of graft incompatibility symptoms in Douglas fir. Silvae Genetica 19: 101-107.

_____. 1978. Isoenzyme activities differ in compatible and incompatible Douglas fir graft unions. Forest Sci. 24: 297-303.

_____. 1980. Anatomical symptoms of graft incompatibility in *Pinus monticola* and *P. ponderosa*. Silvae Genetica 29: 77-82.

Deloire, A. and C. Hebant. 1982. Peroxidase and lignification at the interface between stock and scion of compatible and incompatible grafts of *Capsicum* and *Lycopersicon*. Ann. Bot. 49: 887-891.

Eames, A. J. and L. G. Cox. 1945. A remarkable treefall and an unusual type of graft-union failure. Amer. J. Bot. 32: 311-335.

Earle, E. D. 1968. Induction of xylem elements in isolated *Coleus* pith. Amer. J. Bot. 55: 302-305.

Fletcher, W. E. 1964. Peach bud graft union on *Prunus besseyi*. Proc. Int. Plant Prop. Soc. 14: 265-271.

Fosket, D. E. and L. W. Roberts. 1964. Induction of wound vessel differentiation in isolated *Coleus* stem segments *in vitro*. Amer. J. Bot. 51: 19-25.

Gur. A. 1972. Chemical control of pear-quince graft incompatibility. Proc. Symp. Pear Growing, p. 253-264.

_____ and A. Blum. 1973. The role of cyanogenic glycoside in incompatibility between peach scions and almond rootstocks. Hort. Res. 13: 1-10.

_____, R. M. Samish, and E. Lifschitz. 1968. The role of the cyanogenic glycoside of the quince in the incompatibility between pear cultivars and quince rootstocks. Hort. Res. 8: 113-134.

_____, D. Zamet, and E. Arad. 1978. A pear rootstock trial in Israel. Sci. Hort. 8: 249-264.

Hardham, A. R. and M. E. McCully. 1982a. Reprogramming of cells following wounding in pea (*Pisum sativum*) roots. I. Cell division and differentiation of new vascular elements. Protoplasma 112: 143-151.

_____ and _____. 1982b. _____. II. The effects of caffeine and colchicine on the development of new vascular elements. Protoplasma 112: 152-166.

Harrison, M. A. and R. M. Klein. 1979. Role of growth regulators in initiation of secondary xylem and phloem cells. Bot. Gaz. 140: 20-24.

Hartmann, H. T. and D. E. Kester. 1975. Plant propagation: principles and practices. Third edition, Prentice-Hall, Inc., Englewood Cliffs, N. J.

Heppner, M. and R. McCallum. 1927. Grafting affinities with special reference to plums. Calif. Agr. Exp. Sta. Bul. 438.

Herrero, J. 1951. Studies of compatible and incompatible graft combinations with special reference to hardy fruit trees. J. Hort. Sci. 26: 186-237.

————— and M. Tabuenca. 1969. Incompatibility between stock and scion. X. Behavior of the peach/myrobolan combination when grafted at the cotyledon stage. Ann. Aula Dei 10:397.

Iversen, T. - H. and T. Aasheim. 1970. Decarboxylation and transport of auxin in segments of sunflower and cabbage roots. Planta 93: 354-362.

Jacobs, W. P. 1952. The role of auxin in differentiation of xylem around a wound. Amer. J. Bot. 39: 301-309.

—————. 1979. Plant hormones and plant development. Cambridge University Press, London.

Kruyt, W. 1847. The effect of growth substances, vitamins, traumatic acid, ethylene-chlorbydrin and warm water on the grafting of some ornamental plants. Nederland Dendrol. Vereen., Yaarboek 16: 83-109.

Lang, A., M. Chailakhyan, and I. Frovola. 1977. Promotion and inhibition of flower formation in a dayneutral plant in grafts with a short-day plant and a long-day plant. Proc. Natl. Acad. Sci. 74: 2412-2416.

Lipetz, J. 1970. Wound-healing in higher plants. Int. Rev. Cytol. 27: 1-28.

Mahlsted, J. P. and E. S. Haber. 1957. Plant propagation. John Wiley and Sons, New York.

McClintock, J. 1948. A study of uncongeniality between peaches as scions and the Marianna plum as a stock. J. Agric. Res. 77: 253-260.

Moore, R. 1981a. Graft compatibility-incompatibility in higher plants. What's New In Plant Physiol. 12: 13-16.

—————. 1981b. Graft compatibility and incompatibility in higher plants. Dev. Compar. Immunol. 5: 377-389.

—————. 1982. Graft development in *Kalanchoe blossfeldiana*. J. Exp. Bot. 33: 533-540.

—————. In Press a. Studies of vegetative compatibility-incompatibility in higher plants. V. A morphometric analysis of the development of a compatible and an incompatible graft. Can. J. Bot.

—————. In Press b. —————. IV. The development of tensile strength in a compatible and incompatible graft. Amer. J. Bot.

————— and D. B. Walker. 1981a. —————. I. A structural study of a compatible autograft in *Sedum telephoides* (Crassulaceae). Amer. J. Bot. 68: 820-830.

————— and —————. 1981b. —————. II. A structural study of an incompatible heterograft between *Sedum telephoides* (Crassulaceae) and *Solanum pennellii* (Solanaceae). Amer. J. Bot. 68: 831-842.

————— and —————. 1981c. —————. III. The involvement of acid phosphatase in the lethal cell senescence associated with an incompatible heterograft. Protoplasma 109: 317-334.

————— and —————. In Press. —————. VI. Grafting of *Sedum* and *Solanum* callus tissue *in vitro*. Protoplasma.

Mosse. B. 1962. Graft-incompatibility in fruit trees. Tech. Commun. No. 28, Comm. Bur. Hort. and Plant Crops, East Malling, England.

Muller-Stoll, W. R. 1938. Versuche uker die Verwenbarkeit der B-Indolylessigsaure als verwachsungsforderundes Mittel in der Rebenveredlung. Angew. Bot. 20: 218-238.

Muzik, T. J. 1958. Role of parenchyma cells in graft union in Vanilla orchid. Sci. 127: 82.

Parkinson, M. and M. M. Yeoman. 1982. Graft formation in cultured, explanted internodes. New Phytol. 91: 711-719.

Roberts, L. W. and D. E. Fosket. 1966. Interaction of gibberellic acid and indole-acetic acid in the differentiation of wound vessel members. New Phytol. 65: 5-8.

Sachs, T. 1968. The role of the root in the induction of xylem differentiation in peas. Ann. Bot. 32: 391-399.

Scott, T. K. and W. R. Briggs. 1960. Auxin relationships in the Alaska pea (*Pisum sativum*). Amer. J. Bot. 47: 492-499.

Shimomura, T. and K. Fuzihara. 1976. Histological observations of graft union formation in cactus. J. Japan. Soc. Hort. Sci. 44: 402-408.

———— and ————. 1977. Physiological study of graft union formation in cactus. II. Role of auxin on vascular connection between stock and scion. J. Japan. Soc. Hort. Sci. 45: 397-406.

———— and ————. 1978. Prevention of auxin-induced vascular differentiation in wound callus by surface-to-surface adhesion between calluses of stock and scion in cactus grafts. Plant and Cell Physiol. 19: 877-886.

Stoddard, F. L. and M. E. McCully. 1979. Histology of the development of the graft union in pea roots. Can. J. Bot. 57: 1486-1501.

———— and ————. 1980. Effects of excision of stock and scion organs on the formation of the graft union in *Coleus*: a histological study. Bot. Gaz. 141: 401-412.

Thimann, K. V., T. Sachs, and K. N. Mathur. 1971. The mechanism of apical dominance in *Coleus*. Physiol. Plant. 24: 68-72.

Waisel, Y., I. Noah, and A. Fahn. 1966. Cambial activity in *Eucalyptus camaldulensis* Dehn. New Phytol. 65: 319-324.

Warren Wilson, J. 1978. The position of regenerating cambia: auxin/sucrose ratio and the gradient induction hypothesis. Proc. Roy. Soc. Lond. B 203: x 153-176.

———— and P. Warren Wilson. 1981. The position of cambia regenerating in grafts between stems and abnormally-oriented petioles. Ann. Bot. 47: 473-484.

Yeoman, M. M., D. C. Kilpatrick, M. B. Miedzybrodzka, and A. R. Gould. 1978. Cellular interactions during graft formation in plants, a recognition phenomenon? Symp. Soc. Exp. Biol. 32: 139-160.

Zimmerman, D. C. and C. A. Coudron. 1979. Identification of traumatin, a wound hormone, as 12-Oxo-trans-10-dodecenoic acid. Plant Physiol. 63: 536-541.

VII

Compatability Responses in the Establishment of Mycorrhizae

Edward Hacskaylo

Mycology Laboratory, Plant Protection Institute,
U.S. Dept. Agriculture, Agricultural Research Service,
Beltsville Agricultural Research Center, Beltsville, Maryland 20705

INTRODUCTION

Research on mycorrhizae began about a century ago. I find it fascinating that many early hypotheses on functions of mycorrhizae based primarily on morphological observations were later shown to be very accurate. Numerous studies in progress today are extensions of recorded observations and hypotheses published between 1885 and 1945. Since several excellent references are now or soon will be available (Harley, 1969; Sanders et al., 1975; Marks and Kozlowski, 1973; Schenck, 1982; Harley and Smith, 1983), I shall not provide a general review of mycorrhizae. Rather, I shall discuss selected physiological and morphological data that relate to compatability responses primarily in establishment of ectomycorrhizal associations.

Strictly defined, mycorrhizae are composite structures comprised of specialized symbiotic fungi and the smallest orders of lateral roots of spermatophytes. The term is frequently used to include associations of symbiotic fungi within rhizomes, thalli and even orchid seeds, but I shall discuss only mycorrhizal associations involving true roots.

All mycorrhizal associations may be considered as cases of reciprocal parasitisms, commonly designated as symbiosis. There have been a few reports that imply pathogenic tendencies of certain mycorrhizal fungi when in stress situations, but overwhelming evidence supports their non-pathogenic nature. Some fungi, particularly *Rhizoctonia* spp. (Burgeff,

1909; Downie, 1957; Harvais and Hadley, 1967) and *Armillaria mellea* (Vahl) Fr. (Kusano, 1911; Hamada, 1940), which are pathogens to some plants, may become symbionts with orchids.

Formation of mycorrhizae is usually initiated from soil propagules when lateral roots emerge from parent roots, although colonization may occur later in the growing root tip. In ectomycorrhizae, the fungus penetrates the lateral root in the region of elongation prior to cell maturation. In all types of mycorrhizae the colonization may occur later in the growing root tip. In ectomycorrhizae, the fungus penetrates the lateral root in the region of elongation prior to cell maturation. In all types of mycorrhizae the colonizing fungus invades the cortex but not the stele or apical meristem. Ectomycorrhizal hosts are generally within the Coniferae, Betulaceae and Fagaceae. Most other plants are colonized by endomycorrhizal fungi. Ectendomycorrhizae occur on some ectomycorrhizal hosts and combine morphological features of ecto- and endomycorrhizae. It is common for a single host to be associated with several species of mycorrhizal fungi. Three families of angiosperms (i.e., Chenopodiaceae, Cruciferae and Cyperaceae are generally free of mycorrhizal associations.

Early studies by Melin (1925, 1927) and his co-workers established certain physiological and environmental requirements for many ectomycorrhizal fungi. They tolerate a fairly wide soil acidity, but prefer the pH 4.5-6.0 range. Some, such as *Pisolithus tinctorius* (Pers.) Coker & Couch, can tolerate low levels around pH 3.0 (Marx, 1971), whereas truffle fungi, *Tuber* spp., can grow in calcareous soils at pH 7.0 or above. Ectomycorrhizal fungi are aerobic and prefer a medium that permits a good exchange of oxygen and carbon dioxide. They will not tolerate soils that are saturated with water, primarily because the oxygen tension in those soils is low. Drought and temperature tolerance varies among species (Worley and Hacskaylo, 1959; Mexal and Reid, 1973). Although most appear to grow best at temperatures 20° - 26°C, a few such as *P. tinctorius* will tolerate temperatures above 30°C (Marx et al., 1970). Ectomycorrhizal fungi that have been cultured are thiamine deficient. However, many are fastidious, and complete culture requirements of these ectomycorrhizal fungi are unknown. Ammonium nitrogen sources are preferred over nitrates. With certain exceptions, ectomycorrhizal fungi are unable to hydrolyze most complex carbohydrates. Within genera, some species appear to be faculative symbionts (Hacskaylo and Bruchet, 1972).

ROLE OF ROOT METABOLITES
IN ECTOMYCORRHIZAE

M-Factor

Spores of many species of ectomycorrhizal fungi do not readily germinate on laboratory media. Melin (1954), however, discovered that excised roots of *Pinus sylvestris* L. had a highly stimulatory effect upon germination of spores of certain ectomycorrhizal fungi and upon growth of their mycelia in vitro. He used the designation "M-factor" for what was assumed to be an unidentified root metabolite(s). The M-factor was (1) evident only in the presence of living roots, (2) diffusable through a cellophane membrane, and (3) increased mycelial growth manyfold. Although this discovery was made 30 years ago, Melin was unable to identify the M-factor, and its identity remains a mystery. Recently Birraux and Fries (1981) found that root exudates from *Pinus sylvestris* promoted germination of spores of *Thelephora terrestris,* a mycorrhizal fungus that I worked with for several years but was unable to culture from spores. Their work refocuses our attention to the role and chemical nature of the M-factor in establishing and maintaining ectomycorrhizae.

Carbon Compounds

We can assume that carbon compounds available to the fungus at the root surface are in concentrations sufficient to support colonization of receptive roots. Björkman (1942) hypothesized that since mycorrhizal fungi can generally assimilate only soluble carbohydrates, the absence or presence of sugars within the roots could be directly correlated with formation of mycorrhizae. He further stated that mycorrhizal fungi will not enter the roots unless the roots contain a certain quantity of what he called "surplus soluble carbohydrates." He illustrated this by binding stems of 3-year-old *Pinus sylvestris* plants with a thin wire so that translocation from needles to the roots was interrupted. Under these conditions there was very rapid cessation of mycorrhizal formation. However, plants that were not strangulated continued to develop mycorrhizae. When the wires were removed and fresh conducting tissues developed, the ability to form mycorrhizae was restored. By analyzing tissues for reducing substances he correlated the quantities of soluble carbohydrates in the roots with the frequency of mycorrhizae.

Analyses of carbohydrates in roots showed that the sugar content was low in heavily shaded plants. Sugar content was also low in plants grown at high light intensity and high levels of nitrogen and phosphorous. Björkman concluded that when there is an extreme deficiency or a high sufficiency of nutrients, such as nitrogen and phosphorus, the sugar concentration is reduced in the plant. An extreme deficiency of nutrients reduces plant metabolism, including carbohydrate formation, whereas, high sufficiency favors rapid conversion of carbohydrates into amino acids. Generally, deficiency of phosphorus reduces the concentration of sugar by inhibiting photosynthesis. However, if the deficiency is only moderate, the assimilation of carbohydrate is normal and the formation of protein is reduced. Also, there will occasionally be a surplus of sugar.

In experiments with *Thelephora terrestris* (Ehrh.) Fr. (Hacskaylo, 1965) I demonstrated that sporophore formation of *T. terrestris* occurs very readily in open pot culture in the greenhouse under daylight conditions. However, by removing the shoots from the plants or by covering the plants with a cloth that is impervious to light, sporocarp formation ceases immediately. If the covering is removed, sporocarps resume growth. This extends the concepts of Björkman to link directly survival and reproduction of the fungus to the photosynthetic capability of the host. Further considerations on carbohydrates are discussed later in this paper.

Regulation by the External Nutrient Supply

The external supply of nutrients regulates the development of mycorrhizae, probably by altering the metabolism of the host, thereby changing the availability of metabolites that are essential to the mycorrhizal fungus. High levels of available nitrogen and phosphorus tend to suppress the formation of mycorrhizae. Some mechanism becomes active when the levels of available nutrients are high that limits the amount of available carbohydrates in roots. It is possible that changes in permeability of the plasmalemma of the root cells limits the flow of exudates to the exterior of the root.

Recent papers on endomycorrhizae tend to support this concept. Graham and Menge (1982) grew wheat in phosphorus deficient sandy soil and found a relationship between phosphorus deficiency and increased root exudation. Ratnayake, et al., (1978) demonstrated that phosphorus-induced changes in root exudation altered the phospholipid content of cells, suggesting changes in membrane permeability. A decrease in phospholipids in roots with low phosphorus status resulted in a corresponding increase in membrane permeability and root exudation. The opposite was true in roots having a high phosphorus status. In these roots, high phospholipid levels were positively correlated with a decrease in membrane permeability, and there was less root exudation. Melhuish, et al., (1975) studied the ratios and identities of fatty acids in five mycorrhizal fungi and suggested that lipids might be involved in mycorrhiza formation through changes in membrance permeability.

ROLE OF FUNGAL METABOLITES IN ECTOMYCORRHIZAE

Hormones

If we examine the morphological features of the mycorrhiza of *Pinus virginiana* there is a marked change in its appearance compared to an uninfected root. Root hair development is suppressed. The root tends to be thicker and elongation is reduced. Rather than a single monopodial structure, the root meristem divides dicotymously and produces forked lateral roots, often to the extent that a single short root has the appearance of a cluster of coral, a characteristic only of pine ectomycorrhizae. In studies designed to determine the nature of the stimulus for the changes in root tissues, two particular groups of growth regulators have been implicated: fungus auxin and cytokinins.

Slankis (1949) reported that exudates from *Suillus variegatus* (Swartz ex Fr.) O. Kuntz. produced in axenic culture caused changes similar to mycorrhizae on excised roots of *Pinus sylvestris* L. This morphological alteration could be duplicated by incorporating indoleacetic acid or napthalene acetic acid into the medium. He concluded that fungus auxin was responsible for the general changes in morphology of the pine root during the development of ectomycorrhizae. Other studies on the production of fungus auxin by ectomycorrhizal fungi (Moser, 1959; Ulrich, 1960) relate the production of fungus auxin to the availability of the amino acids tryptophane or alanine and asparagine, from which indole compounds are produced. However, there is very little information available on the identity of fungus auxin. Slankis (1951) reported that the closest thing he could isolate from the medium in which *Boletus edulis* (Bull.) Fr. had grown was indole-proprionic acid. Ulrich (1960) reported that *Suillus variegatus* and *S. granulatus* (L. ex Fr.) O. Kuntze efficiently synthesized indoleacetic acid in a nutrient medium containing only malt extract, glucose, and mineral salts. However, tryptophane supplement was necessary for the production of IAA by other symbiotic fungi that were grown in the same medium. Strzelczyk, et al., (1977) found that fungi isolated from pine mycorrhizae could synthesize auxin from media containing indole, anthranilic acid or indole plus serine. We have performed bioassays using soybean callus tissue in attempts to determine which of approximately 60 isolates of ectomycorrhizal fungi produce auxin in vitro (unpublished). Our results were not very consistent. We were unable to demonstrate auxin production by many of the fungal isolates. Others have reported that detection of the production of auxin by ectomycorrhizal fungi is not always possible using bioassays or even by attempts at chemical identification. I would assume, however, that auxin is an integral part of the process of establishing ectomycorrhizae. It may originate within root tissue and also be synthesized in small but continuous quantities at the root surface by ectomycorrhizal fungi using metabolites that are produced by and secreted to the external cell surfaces of the root. Modifications in development of the root (i.e., the absence of root hairs, dichotymous branching, and changes in the orientation of the cortex cells) may be attributed to the auxin.

Carlos Miller (1967), using soybean callus tissue cultures that require the presence of cytokinin for growth, attained growth stimulation of the callus when it was placed on agar in the vicinity of pieces of mycelial inoculum of *Rhizopogon roseolus* (Cda.) Th. Fr. He isolated zeatin and the ribonucleoside of zeatin, and detected a third, and minor, cytokinin (probably zeatin ribonucleotide from the liquid culture media upon which *R. roseolus* was grown). Of seven mycorrhizal fungi examined, five released such compounds but two did not. We have assayed the isolates in our collection for production of cytokinins and have found that only a small number can definitely be identified as cytokinin producers (unpublished). Although the identification of cytokinins is not consistently possible using bioassays or by analyzing the culture media upon which they are grown, one wonders if this plant hormone also might be released by some fungi when in association with the host root.

Although several years have elapsed since many of these findings were originally published, progress in this important area is at a virtual standstill. I believe hormonal relationships are important in establishing and maintaining mycorrhizal associations. Some possible roles of hormones in mycorrhizal establishment and maintenance might be: (1) changes in root morphology, potentially increasing the root surface absorption area by inducing branching and stimulating production of lateral roots; (2) extension of plasticity of the cell walls of the cortex over a longer period of time, resulting in isodiametric rather than elongate configurations of the cortical cells; (3) invasion of the middle lamellae of cortical cells may be possible because auxins and/or cytokinins affect the chemical nature of the middle lamellae, thus allowing for increased fungal penetration; (4) delay of maturation and senescence of the cortical cells, resulting in greater longevity of the mycorrhizal root as compared to the nonmycorrhizal root; and, (5) changes in membrane permeability with potentially marked influences upon movement of mineral nutrients and organic metabolites between fungus and host. Additional research will be necessary to determine if hormones are essential components in establishing and maintaining this very close relationship between two organisms so widely separated in the evolutionary scheme.

Antibiotics

Many ectomycorrhizal fungi produce antibiotic compounds which may inhibit the competitive advances of other microorganisms in the soil. Marx and Davey (1967) identified diatretyne nitrile as a fungus product of *Leucopaxillus cerealis* var. *piceina* (Peck) ined. that is very inhibitory to *Phytophthora cinnamomi* Rands, a root pathogen. Other antibiotics have been identified and have similar properties, but their role(s) in natural situations has not been extensively examined. This might be of more importance than we have assumed in the past. It is possible that the physical barrier that is established when ectomycorrhizal fungi encompass roots is of more consequence in deterring root pathogens than are the antibiotics that might be produced.

ROOT-FUNGUS INTERACTIONS

When the lateral root emerges from the mother root it is rapidly covered by mycelia of the ectomycorrhizal fungus. The meristem is not invaded, but hyphae penetrate the middle lamellae and grow between the walls of elongating cells of the cortex. Mycelia do not penetrate beyond the endodermis. Most accounts that describe formation of an ectomycorrhizae indicate that the fungus first forms a compact mycelial mantle over the root surface. Hyphae then dissolve the middle lamellae by secreting enzymes that permit invasion of the intercellular regions of the cortex. Hyphae spread through these intercellular regions and form a compact intercellular layer that is designated as the Hartig net.

Recently, however, Nylund (1981) studied the process of ectomycorrhizal formation and observed that hyphae of *Piloderma croceum* Erickss. & Hjortst. which penetrate the intercellular regions of the cortex or short roots of *Picea abies* (L.) Karst. and *Pinus sylvestris* L. assume a labyrinthic mode of growth. The Hartig net is derived from this differentiated type of hyphal growth. The mantle subsequently differentiates from a loose weft of hyphae. Nylund observed that there appeared to be little evidence to indicate enzymatic action on the part of the fungus during penetration of the cortex. Earlier, Lindeberg and Lindeberg (1977) and later Worthington, et al., (1981), suggested that enzymatic activity does occur at the hyphal tips, thus permitting dissolution of the middle lamellae. Nylund (1981) also observed that plasmodesmata were retained in the invaded cortical cells. Whether this pattern of development is common with other hosts and fungi is unknown, and should be critically examined.

Regulation of the degree of infection by the host cells is poorly understood. Krupa and Fries (1971) found that there may be a biochemical and physiological reaction of cortical cells to infection by ectomycorrhizal fungi. Following the invasion of the roots of *Pinus sylvestris* by *Suillus variegatus* there was production and accumulation of volatile terpenes and sesquiterpenes in concentrations up to eight times greater than that found in nonmycorrhizal roots. Since many of those compounds are fungistatic they were considered to be produced as a nonspecific response of the host cells to symbiotic infection. Melin (1955) demonstrated the presence in roots of a diffusible substance which could inhibit the development of ectomycorrhizal fungi. However, he did not provide additional information on the possible identity of such a compound. Other barriers limiting the invasion by hyphae have been suggested. Wilcox (1968) noted that an impervious metacutinized layer is produced over the lateral root apex at the end of the growing season, and that the fungi cannot penetrate this barrier except where there are "breaks." There have also been reports of phenolic compounds being produced by roots in response to invasion by mycorrhizal fungi. These phenolic compounds may also restrict the penetration of fungi to the intercellular regions. Finally, the suberized layers of the endodermis appear to be an effective barrier against movement of the fungus into the stele. Many questions remain regarding the means of invasion, the involvement of enzymes, and the mechanisms for retaining the balance between the host and its endophyte in limiting the extent of the invasion.

As I mentioned previously, Björkman hypothesized that formation of mycorrhizae was dependent upon the availability of soluble sugars within the susceptible roots. Other investigators have attempted to clarify conditions that are required for establishment and maintenance of mycorrhizae. Some of these investigations are in disagreement with the hypothesis of Björkman (Handley and Sanders, 1962; Schweers and Meyer, 1970). Our experiments have shown that high levels of nitrogen and phosphorus or low light intensity suppress formation of mycorrhizae in pine. We also found that short photoperiods suppress ectomycorrhiza formation, probably as the result of a limited amount of photosynthate accumulating within the plant and being translocated to the root. An extensive study of absorption and metabolic pathways of carbohydrates in roots of an ectomycorrhizal host, *Fagus sylvatica* L., was performed by Lewis and Harley (1965a, 1965b, 1965c). By analyzing mycorrhizae and nonmycorrhizal roots of beech, they found that trehalose and manitol were present in the mycorrhizae but not in uninfected roots. They concluded that ectomycorrhizal fungi obtained these sugars from the host through a concentration gradient which is established and maintained by the conversion of sugars in the hyphae into forms that the host will use poorly (i.e., trehalose) or not at all (i.e., manitol). The principal carbon compounds involved in metabolism of the fungus within the root apparently were sucrose, glucose and fructose. By stripping the mycelial sheath from the core of the mycorrhizae and then studying the absorption of carbohydrates, they found that the core responded in a manner similar to uninfected roots. That is, the core contained sucrose.

The sheath, however, contained only trehalose, manitol and insoluble glycogen. Melin and Nilsson (1957) demonstrated that carbon moves from seedlings of *Pinus sylvestris* into associated mycorrhizal fungi. Lewis and Harley concluded that translocation of sugars from the root into the fungal sheath tends to conserve a supply of carbohydrates for the metabolic processes of the fungi. The fungal sheath, hence, acts as a physiological sink, and thus restricts the movement of sugars back into the host. However, there are some indications that metabolites may move from the fungus back into the host (Reid, 1971). Also, Björkman (1960) found that labelled carbon could be translocated from *Picea* and *Pinus* through fungal hyphae to *Monotropa* plants.

Harley and Jennings (1958) suggested that there was external hydrolysis by surface invertase on roots of beech. Palmer and Hacskaylo (1970) were unable to show utilization of sucrose as a carbon source for growth in axenic culture and also believe that hydrolysis occurs at the root surface. Effective utilization of sucrose by the fungal sheath may require the induction of an appropriate adaptive enzyme. France and Reid (1982) suggested that the carbon sink in the fungal sheath could be established by utilizing carbohydrates for growth.

Carbohydrates produced by roots probably are the major, if not the entire, sources for assimilation of and respiration by the entire thallus of the fungus. Without carbohydrates the fungus would be unable to complete its life cycle. If there is an interruption of the availability of carbohydrates to the fungus, the fungus will cease to remain as an associate of the root. The new roots that are formed will be free of mycorrhizae, and fruiting of the fungus will not take place. There is a limiting gap in our knowledge of the transformations of carbohydrates at the interface between the host and fungus and of actual mechanisms involved in the transfer of metabolites from one organism to the other. If better understood, many of the symbiotic relationships in the plant world might be clarified.

Uptake and translocation of nutrients by ectomycorrhizal fungi greatly exceeds that of a root that does not have an associated mycorrhizal fungus. Kramer and Wilbur (1949) first demonstrated that uptake of phosphorus in pine is greater in mycorrhizal roots than in nonmycorrhizal roots. Shortly thereafter, Melin and Nilsson (1950, 1952, 1955) performed a series of experiments on *Pinus sylvestris* with radioactive phosphorus, nitrogen, calcium, and demonstrated that uptake and movement of these nutrients into the host plant were much greater in plants having mycorrhizal fungi associated with roots. They also performed experiments with intact and decapitated plants (1958), and were of the opinion that transpiration was essential for the movement of the phosphorus through the mycorrhizal system and up into the host. Harley and his associates (see Harley, 1969) demonstrated that movement of phosphate from the fungus into the host plant could be prevented by metabolic inhibitors. This indicated that movement of phosphorus involves more than movement by diffusion or movement through the transpiration system. Harley and McCready (1952) found in studies with excised roots of *Fagus sylvatica* that approximately 90% of the phosphate remained in the mycorrhizal sheath as a pool and did not mix with incoming phosphate. Furthermore, phosphate stored in the sheath was released in the plant during periods when external supplies of phosphate were removed. Hence, the sheath appeared to serve as a reserve in times of stress, but was a storage area when phosphates were readily available. Bowen (1973) found that phosphate moves from the Hartig net into the endodermis when hyphae are in contact with the endodermis. However, if the Hartig net does not extend to the endodermis, movement of phosphate is first through the cortical cells and then into the endodermis and stele. There are no data to indicate storage of other nutrients in the mycorrhizal sheath. Most nutrients are translocated directly from the Hartig net into the cortical cells or into the endodermis as they are absorbed from the soil. Many questions remain regarding the identities of compounds active in nutrient uptake, as well as the transfer of nutrients within the fungus and root.

ENDOMYCORRHIZAE

Endomycorrhizal associations are becoming increasingly important as more endomycorrhizal fungi are identified and their physiologies are studied. Endomycorrhizae appear to remain obligate parasites and are not readily cultured. Therefore, physiological studies are difficult to perform. Through manipulation of spores it is possible to work with the isolated organism (to some extent) and in combination with a host. The major element studied in association with uptake mechanisms by endomycorrhizal fungi is phosphorus. Endomycorrhizal associations appear to be particularly advantageous in phosphorus deficient soils. Undoubtedly, uptake of other nutrients is also increased as in ectomycorrhizal associations.

There is an intimate association of fungal hyphae with the protoplasm of cortical cells in endomycorrhizae. The fungus invades the interior of cortical cells, and hyphae are in contact with the plasmalemma. Arbuscules, the small treelike branches at the tips of internal hyphae, are apparently the sites of release of phosphorus into the host tissue. The fungus is subsequently digested by the host cell. Since endomycorrhizae are found on most plants (other than the few groups that I mentioned as being ectomycorrhizal or nonmycorrhizal), their potential in agriculture could become very great.

Since it is inevitable that there are going to be attempts to manipulate the genomes of mycorrhizal fungi, all of the elements required for the symbiotic state will have to be maintained in the altered organism. However, only by intensively researching these organisms can we clarify which elements would be protected in the final product.

POTENTIAL OF GENETIC MODIFICATIONS OF MYCORRHIZAL FUNGI

Giles and Whitehead (1975, 1976, 1977a, 1977b) reported that the spheroplasts of a *Azotobacter vinelandii* were successfully fused with those of the ectomycorrhizal fungus, *Rhizopogon roseoleous*. Their objective was to incorporate the N-fixing property of the bacterium into the hyphae of *R. roseoleous*. This was accomplished, as evidenced by ethylene reduction. However, the authors noted a tendency for one strain of the fungus to become pathogenic toward the roots of *Pinus radiata* D. Don during attempts to synthesize ectomycorrhizae. I am aware of only two preliminary experiments now in progress in the United States directed toward incorporating N-fixing capabilities into ecto- and endomycorrhizal fungi. I hope others are in progress because, I believe, this may be an elegant approach to incorporating nitrogen fixing potential into root systems. If nif-genes could be incorporated into the roots of agricultural crops, nitrogen fertilizers could be conserved and yields of agricultural crops growing on soils of low fertility could be increased. Nitrogen fixation occurs in certain legumes and actinorhizal plants. It has been shown that the growth and yields of peanuts (Daft and El Giahmi, 1976), clovers (Powell, 1976) and black locust (J. Hetrick, personal communication) was greater when individual plants had both endomycorrhizal and N-fixing bacterial associations than when they had either relationship separately. Similar observations have been noted for actinorhizal plants (Daft and Hacskaylo, 1976). Although there are formidable problems involved, the natural compatability of mycorrhizal fungi with root tissues of higher green plants should give priority to modifying mycorrhizal fungi to fix nitrogen over attempts to modify the genetic constituents of root cells. The transfer of nif-genes or incorporation of sphaeroplasts, organalles or DNA from bacteria or Actinomycetes into the true fungi, might be more easily attainable than attempting to incorporate the N-fixing capability of these microorganisms into root tissues. Preference may ultimately be with the Actinomycete, *Frankia*, DNA incorporated into either ecto- or endomycorrhizal fungi, or both. Apparently the anaerobic conditions required by Actinomycetes for N-fixation do not appear to be so stringent as with nodule bacteria and true fungi. Also, Actinomycetes are more closely allied in the evolutionary scheme. The wide host range of endomycorrhizal fungi could be a distinct advantage in using one species or strain as a mycorrhizal associate on a wide variety of important hosts. The impact of these possibilities is self-evident in phosphorus deficient soils were endomycorrhizal fungi greatly increase nutrient uptake. I see no reason to doubt that, in the future, the knowledge and techniques will exist to allow for alteration of the genetic constituents of mycorrhizal fungi. Alterations may include splicing of genetic materials or changes that provide the best of many attributes—tolerance to various ranges in temperature, soil, pH, and moisture levels; increased efficiencies in uptake of nutrients, particularly nitrogen and phosphates; wider ranges of host-fungus compatibility; and even selectivity toward ectomycorrhizal fungi that produce edible sporocarps. Whether any or all of these objectives can be acomplished can only be solved by long-range planning and by intensive efforts by well-trained teams of scientists in many disciplines.

CONCLUSION

I would like to quote what I stated in an article for BioScience 10 years ago (Hacskaylo, 1972), but which is still applicable—"Undeniably, mycorrhizal associations are physiologically among the best examples of balanced reciprocal parasitism in existence. Evolutionary development of mycorrhizal fungi and their hosts has created an interdependency that requires an uninterrupted exchange of certain essential metabolites. Those who grow plants should understand that interruption or imbalance within this relationship, can impair, and eventually destroy, the symbiotic association. Without adequate compensation in the environment the demise of the associated organisms follows."

LITERATURE CITED

Birraux, D., and N. Fries. 1981. Germination of *Thelephora terrestris* basidiospores. Can. J. Bot. 59: 2062-2064.

Björkman, E. 1942. Über die bedingungen der Mykorrhizabildung bei Kiefer und Fichte. Symb. Bot. Upsal. 6: 1-190.

————. 1960. *Monotropa hypopitys* L.—An epiparasite on tree roots. Physiol. Plant. 13: 308-327.

Bowen, G. D. 1973. Mineral nutrition of ectomycorrhizae. *In*: G. C. Marks and T. T. Kozlowski (eds.). Ectomycorrhizae, pp. 151-205. Academic Press, Inc., New York.

Burgeff, H. 1909. Die Wurzelpilze der Orchideen, ihre Kultur und ihr Leben in der Pflanze. G. Fischer, Jena. 220 pp.

Daft, M. J., and A. A. El-Giahmi. 1976. Studies on nodulated and mycorrhizal peanuts. Ann. Appl. Biol. 83: 273-276.

————, and E. Hacskaylo. 1976. Arbuscular mycorrhizas in the anthracite and bituminous coal wastes of Pennsylvania. J. Applied Ecology 13: 523-531.

Downie, D. G. 1957. *Corticium solani*—an orchid endophyte. Nature 179: 160.

France, R. C., and C. P. P. Reid. 1982. Interactions of nitrogen and carbon in the physiology of ectomycorrhizae. Proc. 5th North American Conf. on Mycorrhizae. (In press).

Giles, K. L., and H. Whitehead. 1975. The transfer of nitrogen fixing ability to a eukaryote cell. Cytobios 14: 49-61.

————, and ————. 1976. Uptake and continued metabolic activity of Azotobacter within fungal protoplasts. Science 193: 1125-1126.

————, and ————. 1977a. The localisation of introduced Azotobacter cells within the mycelium of a modified mycorrhiza (*Rhizopogon*) capable of nitrogen fixation. Pl. Sci. Lett. 10: 367-372.

————, and ————. 1977b. Reassociation of a modified mycorrhiza with the host plant roots (*Pinus radiata*) and the transfer of acetylene reduction activity. Pl. and Soil 48: 431-452.

Graham, J. H., and J. A. Menge. 1982. Influence of vesicular-arbuscular mycorrhiza and soil phosphorus on take-all disease of wheat. Phytopath. 72: 95-98.

Hacskaylo, E. 1965. *Thelephora terrestris* and mycorrhizae of Virginia pine. For. Sci. 11: 401-404.

_____. 1972. Mycorrhiza: The ultimate in reciprocal parasitism? Bioscience 22: 577-583.

_____, and G. Bruchet. 1972. Hebelomas as mycorrhizal fungi. Bull. Torr. Bot. Club 99: 17-20.

Hamada, M. 1940. Physiologisch-morphologische Studien über *Armillaria mellea* (Vahl) Quel., Ein Nachtrag zur Mykorrhiza von *Galeola septentrionalis* Reichb. F. Jap. J. Bot. 10: 388-463.

Handley, W. R. C., and C. J. Sanders. 1962. The concentration of easily soluble reducing substances in roots and the formation of ectotrophic mycorrhizal associations—A re-examination of Bjorkman's hypothesis. Pl. and Soil 16(1): 42-61.

Harley, J. L. 1969. The Biology of Mycorrhiza. Leonard Hill Books, 2nd Ed., London. 334 pp.

_____, and D. H. Jennings. 1958. The effect of sugars on the respiratory response of beech mycorrhizas to salts. Proc. Roy. Soc. Ser B, 148: 403-418.

_____, and C. C. McCready. 1952. Uptake of phosphate by excised mycorrhizal root of beech. III. The effect of the fungal sheath on the availability of phosphate to the core. New Phytol. 51: 342-348.

_____, and S. E. Smith. 1983. Mycorrhizal Symbiosis. Academic Press, London. (In press).

Harvais, G., and G. Hadley. 1967. The relation between host and endophyte in orchid mycorrhiza. New Phytol. 66: 205-215.

Kramer, P. J., and K. M. Wilbur. 1949. Absorption of radioactive phosphorus by mycorrhizal roots of pine. Science 110(2844): 8-9.

Krupa, S., and N. Fries. 1971. Studies on ectomycorrhizae of pine. I. Production of volatile organic compounds. Can. J. Bot. 49: 1425-1431.

Kusano, S. 1911. *Gastrodia elata* and its symbiotic association with *Armillaria mellea*. J. Agric. Tokyo 4: 1-66.

Lewis, D. H., and J. L. Harley. 1965a. Carbohydrate physiology of mycorrhizal roots of beech. I. Identity of endogenous sugars and utilization of exogenous sugars. New Phytol. 64: 224-237.

_____, and _____. 1965b. Carbohydrate physiology of mycorrhizal roots of beech. II. Utilization of exogonous sugars by uninfected and mycorrhizal roots. New Phytol. 64: 238-255.

_____, and _____. 1965c. Carbohydrate physiology of mycorrhizal roots of beech. III. Movement of sugars between host and fungus. New Phytol. 64: 256-269.

Lindeberg, G., and M. Lindeberg. 1977. Pectinolytic ability of some mycorrhizal and saprophytic hymenomycetes. Arch. Microbiol. 115: 9-12.

Marks, G. C., and T. T. Kozlowski (Eds.). 1973. Ectomycorrhizae. Academic Press, New York. 444 pp.

Marx, D. H. 1971. Ectomycorrhizae as biological deterrents to pathogenic root infections. *In*: E. Hacskaylo (ed.), Mycorrhizae. USDA For. Serv. Misc. Publ. 1189, pp. 81-96.

Marx, D. H., W. C. Bryan and C. B. Davey. 1970. Influence of temperature on aseptic synthesis of ectomycorrhizae by *Thelephora terrestris* and *Pisolithus tinctorius* on Loblolly pine. For. Sci. 16: 424-431.

Marx, D. H., and C. B. Davey. 1967. Ectotrophic mycorrhizae as deterrents to pathogenic root infections. Nature 213: 1139.

Melhuish, J., R. Dutky, E. Hacskaylo, and G. Bean. 1975. 10-12-octadecadienoic acid in the mycorrhizal fungus *Corticium bicolor*. Phytopathology 65: 836.

Melin, E. 1925. Untersuchungen über die Bedeutung der Baummykorrhiza. Verlag Gustav Fisher, Jena. 152 pp.

_____. 1927. Studien über die Entwicklung der Nädelbaumpflanze in Rahhumus. II. Medded. Stat. Skogsförsökanst. 23: 433-494.

_____. 1954. Growth factor requirements of mycorrhizal fungi of forest trees. Svensk Botanisk Tidskr. 48: 86-94.

_____. 1955. Neuere Untersuchungen über die Mycorrhizapilze der Waldbäume und das Physiologische Wechselspiel Zwischen ihnen und den Wurzeln der Bäume. Uppsala Univ. Årsskr. 3: 1-29.

_____, and H. Nilsson. 1950. Transfer of radioactive phosphorus to pine seedlings by means of mycorrhizal hypae. Physiol. Plant. 3: 88-92.

_____, and _____. 1952. Transport of labelled nitrogen from an ammonium source to pine seedlings through mycorrhizal mycelium. Svensk Botanisk Tidskr. 46: 281-285.

_____, and _____. 1955. Ca^{45} used as indicator of transport of cations to pine seedlings by means of mycorrhizal mycelium. Svensk Botanisk Tidskr. 49: 119-122.

_____, and _____. 1957. Transport of C^{14}-labelled photosynthate to the fungal associate of pine mycorrhiza. Svensk Botanisk Tidskr. 51: 166-186.

_____, and _____. 1958. Translocation of nutritive elements through mycorrhizal mycelia to pine seedlings. Botaniska Notiser 3: 251-256.

Mexal, J., and C. P. P. Reid. 1973. The growth of selected mycorrhizal fungi in response to induced water stress. Can. J. Bot. 51: 1579-1588.

Miller, C. O. 1967. Zeatin and zeatin riboside from a mycorrhizal fungus. Science 157: 1055-1057.

Moser, M. Beiträge zur Kenntnis der Wuchsstoffbeziehungen im Bereich ectotropher Mycorrhizen I. Archiv für Mikrobiologie 34: 251-269.

Nylund, J. 1981. The formation of ectomycorrhiza in conifers: Structural and physiological studies with special reference to the mycobiont, *Piloderma croceum* Erikss. & Hjorst. Acta Universitatis Upsalaliensis. pp. 1-34.

Palmer, J. G., and E. Hacskaylo. 1970. Ectomycorrhizal fungi in pure culture. I. Growth on single carbon sources. Physiol. Plant. 23: 1187-1197.

Powell, C. L. 1976. Mycorrhizal fungi stimulate clover growth in New Zealand hill country soils. Nature 264: 436-438.

Raynayake, M., R. T. Leonard, and J. A. Menge. 1978. Root exudation in relation to supply of phosphorus and its possible relevance to mycorrhizal formation. New Phytol. 81: 543-552.

Reid, C. P. P. 1971. Transport of C^{14}-labelled substances in mycelial strands of *Thelephora terrestris*. *In*: E. Hacskaylo (ed.), Mycorrhizae. USDA For. Serv. Misc. Publ. 1189. pp. 222-227.

Sanders, F. E., B. Mosse, and P. B. Tinker (eds.). 1975. Endomycorrhizas. Academic Press, London. 626 pp.

Schenck, N. C. (ed.). 1982. Methods and Principles of Mycorrhizal Research. Am. Phylopath. Soc., St. Paul, Minn. 244 pp.

Schweers, W., and F. H. Meyer. 1970. Einfluss der Mykorrhiza auf den transport von assimilaten in die Wurzel. Ber. Deutsch. Bot. Ges. 83: 109-119.

Slankis, V. 1949. Wirkung von ß-Indolylessigsaure auf die dicotomische Verzweigung isolierter Wurzeln von *Pinus sylvestris*. Sv. Bot. Tifskr. 43: 603-607.

_____. 1951. Über den Einçluss von ß-Indolylessigsaure und anderen Wuchstoffen auf das Wachstum von Kiefernwurzeln. I. Symb. Bot. Upsal. 11(3): 1-63.

Strzelczyk, E., J. M. Sitek, and S. Kowalski. 1977. Synthesis of auxins from tryptophane and tryptophane-precursors by fungi isolated from mycorrhizae of pine (*Pinus silvestris* L.) Acta Microbiol. Pol. 26: 255-269.

Ulrich, J. M. 1960. Effect of mycorrhizal fungi and auxins on root development of sugar pine seedlings (*Pinus lambertiana*, Dougl.) Physiol. Plant. 13:493-503.

Wilcox, H. E. 1968. Morphological studies of the roots of red pine, *Pinus resinosa*. II. Fungal colonization of roots and the development of mycorrhizae. Amer. J. Bot. 55: 686-700.

Worley, J. F., and E. Hacskaylo. 1959. The effect of available soil moisture on the mycorrhizal association of Virginia pine. For. Sci. 5:267-268.

Worthington, S. J., H. D. Black, and L. B. Coons. 1981. Entry of *Pisolithus tinctorius* hyphae into *Pinus taeda* roots. Can. J. Bot. 59: 2135-2139.

VIII

The Response of Epidermal Cells

Dan B. Walker

Department of Biology, UCLA
Los Angeles, CA 90024

INTRODUCTION

The epidermal layer of plants serves many functions, some of which vary depending on whether the epidermis is located on a stem, leaf, root or reproductive part. Table 1 contains a list of functions that are most commonly ascribed to epidermal cells. In this article I shall single out the protective function for further discussion.

A protective function immediately calls to mind mechanical protection against abrasion by wind, ice or other factors, and one thinks of protection from phytophagous organisms by use of thick cuticles, trichomes and glandular secretions. However, seldom discussed is the protective role of the epidermis in compatibility responses, which are physiological and developmental rather than mechanical problems faced by the plant. The purpose of this article is to discuss the responses of epidermal cells

---

TABLE 1. A list of functions commonly associated with the epidermis.

Prevent water loss by transpiration

Regulate gas exchange with the environment

Glandular secretions for protection or attraction

Absorption (especially in the case of roots)

Light perception in some cases

Mechanical support and control of elongation growth

Protection of internal tissues

---

that are concerned with the phenomenon of tissue compatibility and to relate these responses to the functioning of the plant as a whole.

My intention is to present a perspective on a developing field of study as food for thought for plant biologists, not to review the literature extensively. Literature citaitons have been kept to a minimum in the main text to improve readability. Expanded discussions of certain topics, including references to published reports and to original research findings, are to be found in the footnote section located near the end of the article.

ROLE OF EPIDERMAL CELLS IN COMPATIBLE TISSUE RESPONSES

Compatible responses in somatic plant tissues refer to interactions between different cells that allow the cells to physiologically tolerate each other's presence and that usually lead to harmonious functioning between the cells. Although we take the observation for granted, the various cells in an intact plant can contribute to the common good of the plant only so long as they do not adversely interact physiologically with each other. Consider a laticifer with its often toxic secondary products positioned adjacent to a parenchyma cell. The interaction of the cells is compatible only because the toxic products are normally sequestered within the laticifer. Upon rupture of a laticifer, other living cells in the area experience a dramatic incompatibility response characterized by cellular necrosis and death.

A more familiar example involving somatic cell compatibility is grafting of plant tissues. By grafting I mean any union of cells that previously were not in direct physical contact. Thus, grafting includes wound closures, stem or root unions that are commonly observed in forest trees, man-made unions as in horticultural grafting, and tissue unions that occur during normal plant development (called postgenital or ontogenetic fusions). In all of these examples, cells have come into direct contact and at the very least have tolerated each other's presence. In most instances the cells have also intimately cooperated in subsequent differentiation of the tissues.

Studies on how cells respond during compatible grafting are common for cortical, pith or vascular parenchyma. The process involves wall deposition at the regions of cell contact to affect adhesion of the cells, dedifferentiation of the cells, and often cell proliferation that produces so-called callus cells which interdigitate to fill in the empty spaces. These activities result in a tissue as compact and as firmly bound as the adjacent non-grafted tissues. Indeed, within several days it becomes impossible to detect where the cells have knitted together.[1] Functionally, such a grafting mechanism is an ideal means for joining severed tissues. In a world of environmental perturbation and herbivorous heterotrophs, wounding of plant tissues must often be a daily, if not hourly, occurrence which requires the repair and reconnection of separated cells.

But would one expect epidermal cells to respond in the same way as the aforementioned subepidermal (i.e., cortical, pith, etc.) cells? If epidermal cells readily produced adhesive depositions, dedifferentiated, or proliferated as do subepidermal cells, each time two surfaces of the plant touched for several hours at a time, the surfaces would graft together. Leaves or stems falling against one another would graft at the points of contact. And primordial leaves packed together at the shoot tip would graft into a conical green mass with a loss of function. In short, few plant organs would be capable of assuming their functional morphology and orientation. The rare occurrence of such teratological "fasciations" is testimony to the fact that evolution provided a mechanism to prevent such grafts from occurring routinely and resulting in functional impairments.

The mechanism is that, unlike subepidermal cells, epidermal cells are developmentally incompetent to form graft unions. Hence, epidermal cells fail to adhere, dedifferentiate or divide in random planes of division. While subepidermal cells respond to contact and initiate grafting within a few hours, intact epidermal cells show no grafting responses even after months of intimate contact.[2] Evidently, an early event in the developmental commitment of a cell to become epidermal is attainment of this property of incompetence to graft. Even the protodermal cells at the shoot and root apical meristems must possess this incompetence to graft in order to prevent fasciations from occurring between primordial tissues. Whatever factors trigger the grafting related responses of wall deposition, cellular dedifferentiation and callus production by cell divisions in subepidermal cells are either lacking or ineffective in epidermal cells.

Exceptional Cases of Epidermal Grafting or Dedifferentiation

As has proven true especially in the study of genetics, analyzing the exceptions to a general rule helps clarify the complexity and mechanics of a system. Exceptions to the rule that epidermal cells do not show grafting related responses include postgenital (ontogenetic) tissue fusions, induced epidermal dedifferentiations in selected *in vitro* culture systems, and the aforementioned (if rarely occurring) teratological fasciations. Note that the unions of tree trunks or roots commonly observed in certain forest species are not epidermal phenomena. Such unions appear to take place only between surfaces covered by periderm, that is, older surfaces where the true epidermis has been replaced by a secondary dermal tissue.[3]

Postgenital Tissue Fusions

Postgenital tissue fusions are naturally occurring exceptions to the rule that epidermal cells will not graft. A postgenital fusion is a special category of graft both because epidermal cells dedifferentiate and because the fusion is a predictable event in the ontogeny of the organs involved. The union occurs between particular surfaces at a precise time during development of an organ. The unions of free petal lobes to form a tubular corolla or the unions of free carpels to form a compound pistil are examples where a postgenital fusion may have taken place. However, such unions may also take place by congenital (phylogenetic) fusion which differs from postgenital fusion because a basal meristematic zone forms the united tissues rather than a grafting of epidermal surfaces (Fig. 1). For a more complete discussion of postgenital and congenital fusions, see Cusick, 1966.

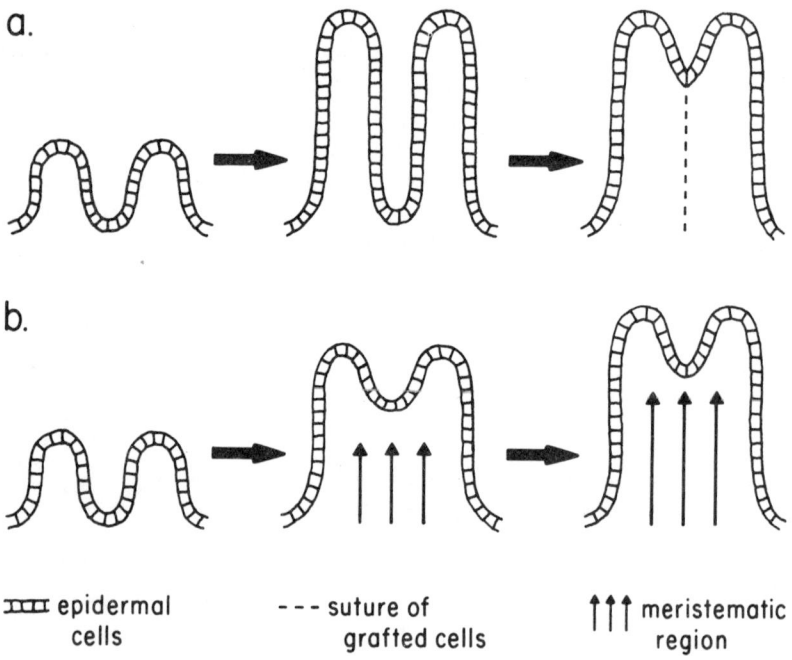

a.

b.

ꞮꞮꞮ epidermal cells   --- suture of grafted cells   ↑↑↑ meristematic region

FIGURE 1. a. A postgenital fusion is the union of free tissue surfaces resulting in the grafting and dedifferentiation of the contacting epidermal cells. b. A congenital fusion is the confluence of tissues resulting from a basal zone of meristematic activity that is continuous between the tissues. Postgenital and congenital fusions can result in united organs of similar morphology at maturity even though their ontogenies are completely dissimilar. Only the postgenital fusion is a true graft between tissues.

During a postgenital fusion, the contacting epidermal cells respond as do successfully grafting subepidermal cells, producing cell wall depositions to cause cellular adherence, rapidly dedifferentiating, and dividing in more random planes of division.[4] Epidermal cells involved in postgenital fusions are developmentally unique among epidermal cells in that they are specially preprogrammed in time and space to be competent to graft. In fact the rate of grafting of these special epidermal cells is more rapid than even that of normal subepidermal cells.[5] The difference in rates of grafting and the spatial and temporal restrictions on grafting in postgenitally fusing epidermal cells suggest that the stimuli for initiating the grafting response are different in postgenital fusions compared to grafts between subepidermal cells. Special forms of cell to cell communication must exist during postgenital fusions that override the inherent incompetence of the epidermal cells to graft.

The functional significance of postgenital unions relates to the evolutionary significance of the fused organs that result. The most numerous examples of fused organs, by either postgenital or congenital means, are petals and carpels. Both petals and carpels are floral parts closely associated with reproductive success. The exploitation of animals, especially insects, to accomplish pollen transfer from anther to stigma led to complex floral morphologies such as landing platforms or attractive beacons for the visitors. The petals, in particular, serve these functions by adaptations in color, shape and symmetry. Differential pollination of a species by agents with long tongues or beaks or with other specialized morphologies led to tubular corollas or to those constructed with special landing platforms. Fusion of individual floral organs was the mechanism to accomplish these specializations in most cases.

The fusion of free carpels to form a compound pistil is probably related to maximizing the seed set in a flower (Carr and Carr, 1961; Walker, 1978b). If the carpels in a flower are fused, pollen deposited on only one stigmatic surface has access to all of the ovules in the flower. In flowers with independent carpels, however, each stigma must independently receive adequate pollen to produce the maximum number of seeds.

Animal pollination evolved rather recently, for the most part concurrent with the early evolution of angiosperms. However, the need to prevent teratological fasciations must have begun with the evolution of the first thalloid or even filamentous alga. It is therefore reasonable to suspect that the incompetence to graft on the part of epidermal cells has long been established in plants. Postgenital unions of epidermal cells must then be more recent evolutionary modifications in plant tissues, where developmental "overrides" reverse incompetence to graft in certain epidermal cells destined for a postgenital tissue union.

Epidermal Dedifferentiation *in vitro*

Epidermal cells have been observed to dedifferentiate and form callus, apical meristems or embryoids on explants cultured *in vitro*. Most recently Tran Thanh Van and coworkers have reported on this phenomenon in what they describe as "thin cell layers" (see review by Tran Thanh Van, 1981). The explants used in these studies are strips of internodal surfaces consisting of the epidermis plus several attached layers of subepidermal (i.e., cortical) cells. Cultured under appropriate conditions, patches of epidermal cells on the explants dedifferentiate to become centers of meristematic activity. However, these reports and others similar to them have in common that the plant was no longer intact, and only isolated pieces of tissue were being cultured. In general, the more intact the tissues of the explant, the less likely will epidermal cells dedifferentiate.[6]

Based on the available research evidence, I propose that morphogenetic factors emanating from tissues internal to the epidermis (possibly vascular tissues) control the incompetence to graft (or dedifferentiate) in the cells of the epidermis. Elimination of these factors by culturing explants which lack vascular tissue would thus allow epidermal dedifferentiation (by loss of incompetence), as has been observed in the "thin cell layer" system. This hypothesis of intertissue communications is currently being explored in more detail in my laboratory.

The functional significance of intertissue communications might be to coordinate the spatial pattern of cell differentiation in plant organs. A model proposing direct communication between different cells or tissues is a departure from the more familiar model of positional cell differentiation mediated by overlapping gradients or fluxes of phytohormones such as auxin (see Sachs, 1981). In the gradient model each cell independently perceives its position by detecting concentrations or polar fluxes of the known phytohormones. By contrast, in a model based on intertissue communications, cells mutually influence one another directly by a local exchange of positional information. This information may be target specific morphogens rather than concentration gradients. Space does not permit a critique of these alternative models in this article, but I believe the future will yield an increasing acceptance of direct cell interactions during pattern formation in plants, although not necessarily to the total exclusion of gradient mediated responses. Different tissues may have different systems of developmental control.

Whatever the underlying mechanism of control, the dedifferentiation of epidermal cells in certain cultured explants seems to be related to disruption of normal tissue and organ integrity. This implies that an intact system of tissues has some means of maintaining certain tissues in their differentiated states in order to guarantee normal functioning. Minor disruptions in the system, such as small wounds, lead to wound closures by dedifferentiating subepidermal cells but not to loss of epidermal or vascular tissues which must still function when the wound has healed. Major disruptions in the system may lead to total loss of tissue integrity, which in nature is probably tantamount to death of the tissues, but which in the laboratory is probably responsible for effects such as epidermal dedifferentiation on cultured explants.

Teratological Fasciations

The study of teratologies (i.e., developmental abnormalities) has lost favor in the last several years, although we still understand little about them. One category of teratology is fasciation (or adnation) where a fusion of parts takes place. Fasciations can arise by processes analogous to postgenital or congenital tissue fusions except that they occur in abnormal and often random locations and impair functioning of the tissues and organs involved. In most cases it has not been reported whether the particular fasciation being described arose by union of free surfaces (i.e. postgenital) or by zonal meristematic activity (i.e., congenital). This distinction is crucial for a discussion of the developmental commitment of epidermal cells since only when free surfaces unite do epidermal cells change fate by dedifferentiation.[7]

Cases of epidermal fasciations are another exception to the rule that epidermal cells are incompetent to graft. Somehow, normal epidermal development must be modified so that incompetence to graft is lost. However, because fasciations severely interfere with normal functioning of parts, they must be regarded as developmental mistakes and not reflective of normal epidermal responses.

ROLE OF EPIDERMAL CELLS
IN INCOMPATIBLE TISSUE RESPONSES

The term tissue incompatibility is often used imprecisely, especially with reference to plant grafts, where it may mean anything from toxic rejection of cells to an unsuccessful union between stock and scion for mechanical or environmental reasons. In this discussion, incompatibility will be used more precisely to refer to physiological interactions between cells that ultimately result in an unsuccessful union or association. A more detailed discussion of incompatibility, with particular reference to plant grafts, is presented in an article by R. Moore in this volume. This more restricted usage of the term incompatibility is consistent with that of plant pathologists to describe the reaction of host cells to pathogen infection. Extensive discussions of host-pathogen interactions are presented by J. Aist and by A. Bell in this volume.

The incompatibility response of plant cells appears to be qualitatively similar whether the stimulus results from pathogen attack, toxic chemicals, or biotic interactions. Cells begin producing defensive compounds such as phytoalexins and hydrolytic enzymes which create a chemical barrier to penetration of the tissues by foreign organisms within a few to several hours after stimulation. Fortification of the cell walls with suberin, lignin, or similar compounds may occur simultaneously. A sever incompatibility response results in cellular autolysis and death of the cells by a characteristic sequence of events. Hydrolytic enzymes are released in the cytosol either from the vacuole or directly from their sites of synthesis. Cellular senscence ensues as organelles and other cytoplasmic inclusions lose their integrity, producing a granular or flocculent appearance of the cytoplasm when viewed with the transmission electron microscope. The last cellular component to survive autolysis appears to be the plasma membrane, which maintains turgor in the senescent cell. Turgor loss and cell collapse eventually occur and must signal the loss of plasma membrane integrity. Cell walls in the region of tissue senescence may become soft and jelly-like. If the incompatibility inducing stimulus is localized, the local region of senescing cells usually becomes sealed off by wound periderm to protect unaffected cells in the plant.[8]

The functional role of the incompatibility response is defense of the plant against penetration by foreign organisms. Once inside the epidermis, a pathogen encounters fewer mechanical hindrances since the cell walls are often thinner and intercellular spaces are present. Thus, a rapid chemical response is a major defensive weapon once the epidermal barrier has been breached.

But the first line, and probably major defense, of the plant is the epidermal layer. In addition, the epidermis is probably more efficient than its replacement tissue, the periderm, in other functions, such as gas exchange. A mechanism to preserve the intact epidermis on a plant by preventing epidermal cells from undergoing incompatibility responses except as a last resort would be of value. The epidermal surfaces of a plant are constantly subjected to toxic substances such as the extruded sap from damaged plants or animals, pathogens, excrements and heavy metals. All of these substances will elicit a rapid incompatibility reaction if placed on subepidermal cells. But will epidermal cells react the same as subepidermal cells?

Recent results from my laboratory indicate that incompatibility inducing substances applied to the intact epidermis elicit little or no response. In particular we have studied incompatible graft combinations and tested the responses of subepidermal cells compared to intact epidermal cells. As expected, an incompatible combination of grafted subepidermal cells results in a typical senescence of the cells along the graft interface. However, intact epidermal cells of a normally incompatible graft combination show little or no response for up to several months; they neither graft nor elicit incompatibility.[9]

The mechanism responsible for this lack of response appears to be the cuticle on the surface of intact cells.[10] The varnish-like nature of plant cuticles creates a largely impermeable covering over the epidermal cells which excludes hydrophilic solutions such as plant saps and water that may carry toxic solutes. The role of the cuticle in preventing water loss from the plant is universally recognized. However, the role of the cuticle in preventing water (or solution) entry into the plant also deserves appreciation. Excluding incompatibility inducing solutions from reaching the protoplast is an important defensive function of the cuticle to prevent unnecessary loss of functional epidermal cells by induced cellular senescence. Only if the outer wall of the epidermis is penetrated to create a portal of entry to potential pathogens would it be prudent for the plant to elicit an incompatibility response in which epidermal cells would be lost and replaced by periderm. Thus, the cuticle functions as a first line, mechanical defense of the plant which acts to shelter the second line, chemical defense of incompatibility responses in epidermal cells from being expressed unless required.

Brief mention should be made of a prominent exception to the general rule that the cuticle of intact epidermal cells acts to exclude incompatibility factors. Stigmatic trichomes can respond to incompatible pollen within only a few minutes despite the presence of a cuticle on the trichomes. Stigmatic surfaces are the sites of recognition in plant breeding systems where determinations of self and like species are made. Inappropriate pollen is excluded from the fertilization process. The cellular incompatibility responses in these instances are often different from those discussed in this paper for somatic cells in that they involve lack of pollen germination, arrested pollen tube growth, failure to penetrate the stigmatic surface or depositions of callose. These responses relate to the unique roles of stigmatic cells in recognition and reproduction. For more elaborate treatments of reproductive incompatibility responses in plants, refer to recent reviews by Heslop-Harrison (1975) and Clarke and Knox (1978).

CONCLUDING REMARKS

The unique position in the plant and the roles assumed by epidermal tissue require specialized structural and developmental characteristics. Because epidermal cells alone in the plant must interact directly with the biotic and abiotic agents of the environment, the responses of epidermal cells are different from, and in some cases the opposite of, interior cells, including those immediately adjacent to the epidermis. Evidence from developmental and structure-function studies is beginning to aid our understanding of how cell differentiation and determination are involved in the important epidermal functions of maintaining the integrity of the plant body while at the same time interfacing with the community and environment surrounding the individual plant. These are day to day functions that involve compatibility and incompatibility responses of somatic plant cells, about which we still know relatively little. The time is ripe and the field is open for combined structural, developmental and biochemical approaches to cell differentiation and interactions of plant cells. The results of such studies will have impact on areas from molecular biology to community ecology because of the fundamental importance of cellular compatibility responses.

ACKNOWLEDGMENTS

The author gratefully acknowledges the assistance of David K. Bruck and Judith A. Verbeke in critically reading the manuscript.

FOOTNOTES

[1]Many studies have been performed on graft unions because of their use in the field of horticulture. Until recently, the analysis of graft unions was made using light microscopic techniques only and emphasis, therefore, was laid primarily on features of callus production and the redifferentiation of vascular tissue across the graft interface. Because callus cells are derived from parenchymatous cells, most observations have focused on the proliferating cells in pith or cortex, or in the vicinity of the vascular cambium in woody tissue grafts. These light microscopic studies are reviewed by M. McCully in this volume.

Recent studies by Moore and Walker (1981 a,c; 1982) using electron microscopy and cytochemistry provided more detailed information on the grafting responses of individual cells. These studies documented the deposition of cell wall materials by dictyosomes plus the cytological events of wound induction and subsequent recovery from wounding in grafted cells.

[2]Recent studies in my laboratory of a wide range of species (Table 2) have indicated that intact epidermal cells lack the competence to graft. We bound together young internodes of two branches on the same plant for periods of up to 4 months. Normal expansion in diameter of the internodes flattened the exposed surfaces and forced them into contact under pressure. In no case did epidermal cells show a grafting response. Each time the grafting tape was removed, the epidermal surfaces readily fell apart and no epidermal cells had dedifferentiated.

To test the effect of dedifferentiating subepidermal cells on intact epidermal cells, we bound together two internodes but with the epidermis removed surgically from the surface of one of the internodes. On the surface with the epidermis removed, the subepidermal cells dedifferentiated and divided within a few

TABLE 2. A list of species on which compatible and incompatible grafting experiments are being performed in the author's laboratory.

Dicotyledons:

Catharanthus roseus	(Apocynaceae)
Exacum affine	(Gentianaceae)
Impatiens sultanii	(Balsaminaceae)
Oxalis herrarae	(Oxalidaceae)
Sedum telephoides	(Crassulaceae)
Solanum pennellii	(Solanaceae)

Monocotyledons:

Scindapsus aureus	(Araceae)
Tradescantia spp.	(Commelinaceae)

days and formed wound periderm within one to two weeks. By contrast, no response occurred in the epidermal cells on the intact internodal surface, even after months in contact with the opposing periderm layer. Nor was there any adhesion developed between the surfaces. The attempted grafts fell apart when the grafting tape was removed.

Thus, neither intimate epidermal cell contact, physical pressure, internal-like cellular environment, nor influences of dedifferentiating subepidermal cells could induce a grafting response in epidermal cells under experimental conditions. These data, taken with the observation that epidermal grafts are rare in nature, lead to the conclusion that one aspect of the state of determination of epidermal cells is an incompetence to elicit responses associated with grafting.

[3]Numerous reports can be found of naturally occurring root grafts between woody plants (Bormann, 1962; Keane and May, 1963; Guengerich and Millikin, 1965; Miller and Woods, 1965; Graham and Bormann, 1966; Rao, 1966; Stone et al., 1973), and the natural union of woody tree trunks is a common sight. The important similarity about these unions is that they are all woody, that is, they occur in plants with an actively expanding vascular cambium and periderm. Tissue union is established by proliferations of ray parenchyma tissues in both grafting partners which eventually break through the periderms at the surfaces to form a united parenchymatous mass (Millner, 1932; Rao, 1966). Epidermal cells do not establish the tissue union between the two partners. Indeed, reviews of the literature by this author and independently by another (Jon Keeley, personal communication) have failed to locate any reports of root or shoot grafts between epidermal tissues, excluding postgenital tissue fusions and the haustorial attachments of vascular plant parasites (see discussions of vascular parasites in this volume by J. Kuijt and J. Riopel).

[4]References to earlier light microscopic studies can be found in Cusick, 1966. More recent studies in my laboratory using electron microscopy have determined the cellular responses of postgenitally fusing cells using electron microscopy

133

(Walker, 1975 a,b,c). Adherence of the contacting epidermal cells results from the rapid deposition of a middle lamella-like layer apparently through the activities of dictyosomes and endoplasmic reticulum. The cell wall depositions are concentrated on the contacting (outer tangential) cell surfaces where the depositions migrate through the young cuticular layers to act as the intercellular substance. At the same time the epidermal cells initiate periclinal cell divisions and begin to lose their specialized characteristics (i.e., dedifferentiate). When fusion is complete, the former epidermal cells have completely redifferentiated, and the original line of union between cells may be entirely obscured.

[5]To test how postgenitally fusing cells are triggered to begin the grafting response, we inserted a barrier of gold foil between the epidermal surfaces prior to ontogenetic contact of the organs destined to fuse. The epidermal surfaces became strongly appressed against the metal surface, excluding intercellular spaces and mimicking an internal cellular environment. However, the cells remained epidermal and were not induced to graft or dedifferentiate. These experiments implicated some form of direct intercellular communication between cells destined to unite postgenitally and eliminated physical pressure or cell confinement as the morphogenetic stimuli for grafting and dedifferentiation (Walker, 1978a).

In more recent work we have followed the kinetics of epidermal differentiation and dedifferentiation in our model system, the fusing carpel primordia of *Catharanthus roseus*. One of the most interesting findings from this study was the rapid response time of postgenitally fusing cells. Some of the epidermal cells completely dedifferentiated within only two hours of first contact and all contacting cells had responded dramatically within 10-12 hours of first contact (Verbeke and Walker, manuscript in preparation). A two hour response time is noticeably faster than that observed for dedifferentiating subepidermal cells in grafting situations.

Preliminary results of surgical manipulations of fusing organs indicate that only certain regions of epidermal surfaces are competent to unite by the postgenital fusion mechanism. We have replaced one of the normally fusing surfaces in the *C. roseus* carpel system with epidermal surfaces (carpel, anther, petal) of various ages that normally do not postgenitally fuse. To date, we have obtained epidermal fusion responses only when the surfaces that normally unite are used; other surfaces appear to be incompetent to fuse (Verbeke and Walker, unpublished results). Thus, certain epidermal cells only must be predetermined to respond to a grafting stimulus.

[6]Tran Thanh Van and co-workers have studied "thin cell layers" precisely for the reason that the cells behave differently when removed from association with other tissues. The advantages of "thin cell layers" to them are that such layers probably have minimal amounts of complicating endogenous regulatory substances and should be relatively free of intertissue correlative effects (Tran Thanh Van, et al., 1974). The following example of such intertissue effects was reported by Chlyah (1974). A cultured tissue explant of epidermal cells died. An explant of epidermal plus subepidermal cells (i.e., a "thin cell layer") lived, and many of the epidermal cells dedifferentiated and divided numerous times. An explant of epidermal plus subepidermal plus vascular cells also lived, but the epidermal cells did not dedifferentiate. Thus, in some way different tissues influence the morphogenetic potential of each other.

Epidermal dedifferentiation has been reported by other workers as well but only under conditions of wounding or experimental treatments where the plant tissue was no longer part of an intact and normally functioning system (Bloch, 1935; Bain, 1940; Dehnel, 1960; Tucker, 1972; Thakur et al., 1977).

[7]For a discussion of the types of teratologies in plants and their manner of development, refer to a review by Gorter (1965). More recently, Heslop-Harrison (1972) has considered the functional significance of teratologies and their value to ontogenetic and phylogenetic studies.

To date the best studies of fasciations arising by a postgenital fusion mechanism are the anatomical descriptions of hybrid honeysuckles (*Lonicera* spp. L.) by Vieth and Lamond (1973) and Lamond and Vieth (1975). These workers documented the unions of leaf and stem sufaces and showed that a dedifferentiation of epidermal cells was involved in some instances.

Fasciations have been induced experimentally using growth regulators, other chemicals and viral infections, but in most cases it was not determined whether union was by postgenital or congenital processes (Gorter, 1965). Identification of a clear-cut instance of inducible postgenital fasciation would provide a useful system for future developmental studies of epidermal determination.

[8]Most of our understanding of the cellular defense response variously called incompatibility, autolysis, or hypersensitivity comes from studies of how plant cells react to pathogen invasion. Deverall (1977) and Bell (1981) provide excellent reviews of these phytopathological and related studies. Also, the reader may refer to the articles by A. Bell and J. Aist in this volume.

We have studied the cellular incompatibility response in my laboratory using an incompatible graft between *Solanum pennellii* and *Sedem telephoides*. Ontogenetic, cytochemical and *in vitro* studies indicated that graft incompatibility involves the same types of cellular autolytic responses as occur in cells reacting to pathogen attack (Moore and Walker, 1981 b,c; 1982). However, the focus of these studies was on pith and cortical cells, and no attempt was made to examine epidermal cell responses.

[9]Various heterograft combinations using the species listed in Table 1 were examined, and two types of grafts were constructed: one type had two intact internodal surfaces bound together and another type had one intact surface bound together with one wounded subepidermal surface of the other partner. For most of the species combinations examined, the epidermal cells responded little or not at all to either type of graft for periods of up to a few months (Walker, manuscript in preparation).

[10]The most compelling evidence that the cuticle acts to exclude incompatibility factors from reaching the epidermal protoplast and thus eliciting an incompatibility response again comes from studies of host cell-pathogen interactions. It has been well documented that epidermal cells will undergo a hypersensitive (i.e., incompatible) response once a pathogen penetrates the outer cell wall but not until the wall (and cuticle) has been penetrated (refer to Deverall, 1977, for a complete discussion). Thus, epidermal cells are innately competent to respond to incompatibility just as are subepidermal cells. The most likely possibility for the failure of most epidermal cells to respond to toxic factors on their outer surfaces is that the toxins never make it through the intact surface of the hydrophobic cuticle.

In attempts to elicit incompatibility responses in epidermal cells using our grafting systems, we constructed heterografts as explained previously (see footnote 9) except that we disrupted the intact surface of the epidermis so that toxins could enter. We then examined intact epidermal and subepidermal cells surrounding the site of disruption to see if an incompatibility response had been induced. Both subepidermal cells and epidermal cells showed signs of cellular senescence that were more severe than wounded control surfaces (Walker, manuscript in preparation). These results are consistent with the suggestion that water soluble toxins are excluded from the epidermal protoplast by the hydrophobic cuticle but will elicit incompatibility responses when they enter the protoplast through the other, hydrophilic walls of the cell.

LITERATURE CITED

Bain, H. F. 1940. Origin of adventitious shoots in decapitated cranberry seed-lings. Bot. Gaz. 101: 872-880.

Bell, A. A. 1981. Biochemical mechanisms of disease resistance. Ann. Rev. Plant Physiol. 32: 21-81.

Bloch, R. 1935. Wound healing in *Tradescantia fluminensis*. Vell. Ann. Bot. 49: 651-670.

Bormann, F. H. 1962. Root grafting and non-competitive relationships between trees. *In* T. T. Kozlowski [ed.], Tree growth. Ronald Press, New York. pp. 237-246.

Carr, S. G. M., and D. J. Carr. 1961. The functional significance of syncarpy. Phytomorphology 11: 249-256.

Chlyah, H. 1974. Inter-tissue correlations in organ fragments. Plant Physiol. 54: 341-348.

Clarke, A. E. and R. B. Knox. 1978. Cell recognition in flowering plants. Quart. Rev. Biol. 53: 3-28.

Cusick, F. 1966. On phylogenetic and ontogenetic fusions. *In* E. G. Cutter [ed.], Trends in plant morphogenesis. Longmans, Green, & Co. Ltd., London.

Dehnel, G. S. 1960. Response of stomata to wounding. Bot. Gaz. 122: 124-130.

Deverall, B. J. 1977. Defense mechanisms in plants. Cambridge University Press, Cambridge.

Gorter, C. J. 1965. Origin of fasciation. Encyclop. Plant Physiol. XV (pt. 2): 330-351.

Graham, B. F. Jr., and F. H. Bormann. 1966. Natural root grafts. Bot. Rev. 32: 255-292.

Guengerich, H. W., and D. F. Millikin. 1965. Root grafting, a potential source of error in apple indexing. Plant Dis. Rpt. 49: 39-41.

Heslop-Harrison, J. 1972. A reconsideration of plant teratology. Phyton 4: 19-34.

_____. 1975. Incompatibility and the pollen-stigma interaction. Ann. Rev. Plant Physiol. 26: 403-425.

Keane, F. W. L., and J. May. 1963. Natural root grafting in cherry and spread of cherry twisted-leaf virus. Canad. Plant Dis. Survey 43: 54-60.

Lamond, M., and J. Vieth. 1975. Contribution à la tératologie des Chèvefeuilles et an problème des fusions III. Association de cymules biflores par gamophyllie ontogénique. Can. J. Bot. 53: 1906-1924.

Miller, L., and F. W. Woods. 1965. Root grafting in loblolly pine. Bot. Gaz. 126: 252-255.

Millner, M. E. 1932. Natural grafting in *Hedera helix*. New Phytol. 31: 2-25.

Moore, R., and D. B. Walker. 1981a. Studies of vegetative compatibility-incompatibility in higher plants. I. A structural study of a compatible autograft in *Sedum telephoides* (Crassulaceae). Amer. J. Bot. 68: 820-830.

————. 1981b. Studies of vegetative compatibility-incompatibility in higher plants. II. A structural study of an incompatible heterograft between *Sedum telephoides* (Crassulaceae) and *Solanum pennellii* (Solanaceae). Amer. J. Bot. 68: 831-842.

————. 1981c. Studies of vegetative compatibility-incompatibility in higher plants. III. The involvement of acid phosphatase in the lethal cellular senescence associated with an incompatible heterograft. Protoplasma 109: 317-334.

————. 1982. Studies of vegetative compatibility-incompatibility in higher plants. VI. Grafting of *Sedum* and *Solanum* callus tissue *in vitro*. Protoplasma (in press).

Rao, A. N. 1966. Developmental anatomy of natural root grafts in *Ficus globosa*. Austral. J. Bot. 14: 269-276.

Sachs, T. 1981. The control of the patterned differentiation of vascular tissues. *In* H. W. Woolhouse [ed.], Advances in botanical research. Academic Press, New York. pp. 151-262.

Stone, E. L., J. E. Stone, and R. C. McKittrick. 1973. Root grafting in pine trees. Food and Life Sci. Quart. 6:19-21.

Thakur, S., P. S. Ganaputhy, and B. M. Johri. 1977. *In vitro* shoot bud differentiation from epidermal cells of stem segments in *Bacopa monniera* (Linn.) Pennell. Beitr. Biol. Pflanz. 53: 321-330.

Tran Thanh Van, K. M. 1981. Control of morphogenesis in *in vitro* cultures. Ann. Rev. Plant Physiol. 32: 291-311.

Tran Thanh Van, K. M., H. Chlyah, and A. Chlyah. 1974. Regulation of organogenesis in thin layers of epidermal and subepidermal cells. *In* H. E. Street [ed.], Tissue culture and plant science. Academic Press, New York. pp. 101-104.

Tucker, S. 1972. Wound repair in leaf blades of Magnoliaceous plants (Angiospermae; Magnoliales). ASB Bull. 19: 107.

Vieth, J., and M. Lamond. 1973. Contribution à la tératologie des Chèvefeuilles et au problème des fusions. Can J. Bot. 51: 517-525.

Walker, D. B. 1975a. Postgenital carpel fusion in *Catharanthus roseus* (Apocynaceae). I. Light and scanning electron microscopic study of gynoecial ontogeny. Amer. J. Bot. 62: 457-467.

————. 1975b. Postgenital carpel fusion in *Catharanthus roseus*. II. Fine structure of the epidermis before fusion. Protoplasma 86: 29-41.

————. 1975c. Postgenital carpel fusion in *Catharanthus roseus*. III. Fine structure of the epidermis during and after fusion. Protoplasma 86: 43-63.

————. 1978a. Morphogenetic factors controlling differentiation of epidermal cells in the gynoecium of *Catharanthus roseus*. I. The role of pressure and cell confinement. Planta 142: 181-186.

————. 1978b. Study of postgenital carpel fusion in *Catharanthus roseus* (Apocynaceae). IV. Significance of the fusion. Amer. J. Bot. 65: 119-121.

IX

The Cell Surface
in Plant Recognition Phenomena

Anthony Bacic and Adrienne E. Clarke

Plant Cell Biology Research Centre
School of Botany, University of Melbourne
Parkville, Vic. 3052, Australia

INTRODUCTION

Although plants have no immune system directly equivalent to that of animals, they do have the capacity to defend themselves from potential pathogens (Deverall, 1977), to form specific symbiotic associations (Bauer, 1981) and to select compatible mating partners (Heslop-Harrison, 1975). This implies that plant cells have the capacity to recognize and respond to contact with other cells, either prokaryotes or eukaryotes (Clarke and Knox, 1978). A similar capacity for recognition and response at the cellular level is fundamental to the animal immune system, and it is possible that the underlying mechanisms for cell-cell recognition may be similar in both plants and animals. However, a fundamental difference between plant and animal cells is the presence of a cell wall which encases the plant protoplast. All extracellular signals to the cytoplasm of plants cells are transmitted through this wall, which may itself be actively involved in determining the outcome of cell-cell interactions.

In this paper we compare the basis of "immunity" in plants and animals by outlining the basis of immunity and its evolution in animal systems; we then consider the idea that while the evolution of the cell wall in plants may have prevented the development of an immune system based on circulating cells as occurs in animals, it may itself be part of the "immune equivalent" of plants. In this article we have been selective and have emphasized the cell wall and its potential role in recognition. References to more comprehensive general reviews on plant cell recognition are Heslop-Harrison, 1975; Curtis, 1978; Clarke and Knox, 1979; Clarke and Gleeson, 1981; Ralton and Clarke, 1982. Reviews of particular recognition systems are algal mating (Wiese and Wiese, 1978; Van den Ende, 1981; Evans *et al.*, 1982), yeast mating (Pierce and Ballou, 1982), grafting (Moore, 1981), host-symbiont (Bauer, 1981; Schmidt and Bohlool, 1981) and host pathogen interactions (Kúc, 1976; Bell, 1981; Kosuge, 1981; West, 1981).

IMMUNITY IN ANIMALS

Immunity is the ability of an organism to defend itself against foreign invaders, and implies that the organism has the ability to recognize and defend itself from non-self organisms. For vertebrates there are two main types of immunity, humoral and cell-mediated immunity, both of which are mediated by a class of white blood cells, the lymphocytes. Humoral immunity is the production of specific antibodies or immunoglobulins, to foreign materials or antigens. Humoral immunity is due to B-lymphocytes, which in mammals, originate in the bone marrow. After recognition of the antigen, B-lymphocytes divide to give rise to antibody-secreting cells. The antigen is recognized by its interaction with a specific membrane-bound receptor on the B-lymphocyte which is itself an immunoglobulin, with the same specificity as the antibody which will ultimately be secreted by the progeny cells.

Cell-mediated immunity depends on T-lymphocytes which develop in the thymus. When a particular antigen or foreign invading cell is recognized by membrane-bound receptors on these cells, they are stimulated to produce "killer-T-cells" which directly kill cells having the same surface antigen which stimulated their production. These two major types of lymphocytes result in a highly adaptive, integrated immune system in animals. These cells circulate through the blood stream, the lymphatics and the tissues: when a foreign antigen is recognized, there is an immediate response, either of production of circulating antibodies or killer-T-cells, or both. Both systems are characterized by memory and specificity. For the humoral response, specificity is expressed in the very precise, complementary interaction of a particular antigen and its antibody; memory is expressed in the heightened response to a second challenge of the same but not a different antigen.

The cell-mediated response can be recognized experimentally as the ability of an animal to reject a foreign graft. This is an artificial situation, but it is believed to be an expression of the purpose for which this type of immunity evolved, that is, rejection of cells such as aberrant or malignant self-cells which are recognized as non-self by the immune system. (See texts: Watson, 1977; Roitt, 1980; Inchley, 1981 for details and further references.)

This integrated B and T lymphocyte-based immune system is found only in vertebrates. In higher invertebrates there are equivalents of T-lymphocytes. Allografts are rejected, and there is short-term memory of grafts. There are no circulating antibodies, but lectins, which might act as non-specific agglutinins of foreign microorganisms, are often present in the haemolymph. In the lower invertebrates there is specific allograft rejection but no memory (Hildeman, 1974; Chorney and Cheng, 1980). Apart from these manifestations of specific immunity, the non-specific responses of phagocytosis and encapsulation of foreign particles are common to all animals. The evolutionary relationships of these immune systems are summarized in Table 1. All of the responses depend on the deformable nature of the cells and on the presence of specific plasma membrane receptors.

TABLE I

	PLANTS		COMMON WALL-LESS UNICELLULAR PROGENITOR	ANIMALS		
	HIGHER PLANTS ←	LOWER PLANTS ←	→	LOWER INVERTEBRATES →	HIGHER INVERTEBRATES →	VERTEBRATES
Evolution of "immune" cell recognition systems	Plasma membrane receptors? ←	Plasma membrane receptors		Plasma membrane receptors →	Plasma membrane receptors →	Plasma membrane receptors
	Cell wall-mediated recognition	Cell wall-mediated recognition			+ T-lymphocyte cell-mediated immunity	+ T-lymphocyte cell-mediated immunity
						+ B-lymphocyte humoral immunity
Expression of particular system involved		Specific aggregation of wall-less algal gametes		Species-specific aggregation of associated somatic cells		
	Wall-mediated recognition in host-pathogen interactions	Wall-mediated recognition in yeast mating				
	Allografts tolerated	Allografts tolerated		Allograft rejection	Allograft rejection	Allograft rejection
				Memory component?	Short term immunological memory	Specific immunological memory
						Antibody production
Non-specific primitive defense	← Encapsulation Agglutination →				← Phagocytosis Encapsulation Agglutination →	

THE CELL WALL IN PLANT CELL-CELL INTERACTIONS

The development of the cell wall was an essential evolutionary event determining differences between the animal and plant kingdoms (Gun-

141

ning, 1982). Multi-cellular organisms containing cells with walls are poorly adapted for functions requiring cell motility and deformability. Thus there was no possibility of development of a surveillance system with memory, based on circulating cells originating from a thymus equivalent in plants. It follows that the characteristics of the immune system in animals such as allograft rejection, antibody production and specific immunological memory would not necessarily be expected in plants. It is believed that cell-cell recognition mechanisms in animals are also involved in, and may have evolved from, the contact regulation required during embryonic growth. In plants, there is apparently no requirement for contact regulation during development, so there would be no evolutionary pressure for development of such mechanisms. However, precise cell-cell interactions do occur in plants. These interactions depend on transmission of signals which must traverse the cell wall to gain access to the plasma membrane, where there is presumably transduction of the original or a secondary signal to elicit the cytoplasmic response. The cell wall may merely present a physical barrier to diffusion of an extracellular signal, or it may itself play an active role (see below). In either case, information regarding the nature and organization of cell wall components is essential to understanding the molecular basis of recognition in cell-cell interactions.

The information required to understand a particular interaction is quite precise. For example, a complete understanding of the symbiotic association between soybean and *Rhizobium japonicum* would require knowledge not merely of cell wall structure of root hairs, but of the walls of the relatively few hair cells at the junction of the zone of elongation and the hair zone, which are induced to form hairs by association with the *Rhizobium*, and which are the only cells apparently capable of forming "infection threads" and nodules (Turgeon and Bauer, 1982). Another example is the interaction between the root-rotting pathogen *Phytophthora cinnamomi* and roots of both susceptible and non-susceptible plants. The pathogen invades root tissues at the zone of elongation (Hinch and Weste, 1979). Motile zoospores are attracted chemotactically to this zone, where they then adhere and enyst at the surface of the mucilage layer covering the epidermal cells. For one non-host plant, *Zea mays*, fucosyl residues of the root mucilage are specifically involved in the interaction (Hinch and Clarke, 1980 a). After encystment, germ tubes grow and hyphae penetrate the epidermal cell wall and proceed to grow within the tissue. In susceptible hosts, growth proceeds into the vascular tissue. However, in non-susceptible hosts such as *Zea mays*, growth is arrested within the epidermal layers (J. Hinch, R. Wetherbee and A. Clarke, unpublished observations) Fig. 1. A knowledge of the structure of both the extracellular mucilage and the epidermal cell walls is essential to an understanding of both the mechanism of the initial zoospore-root interaction, and the arrest of hyphal growth in the non-susceptible host.

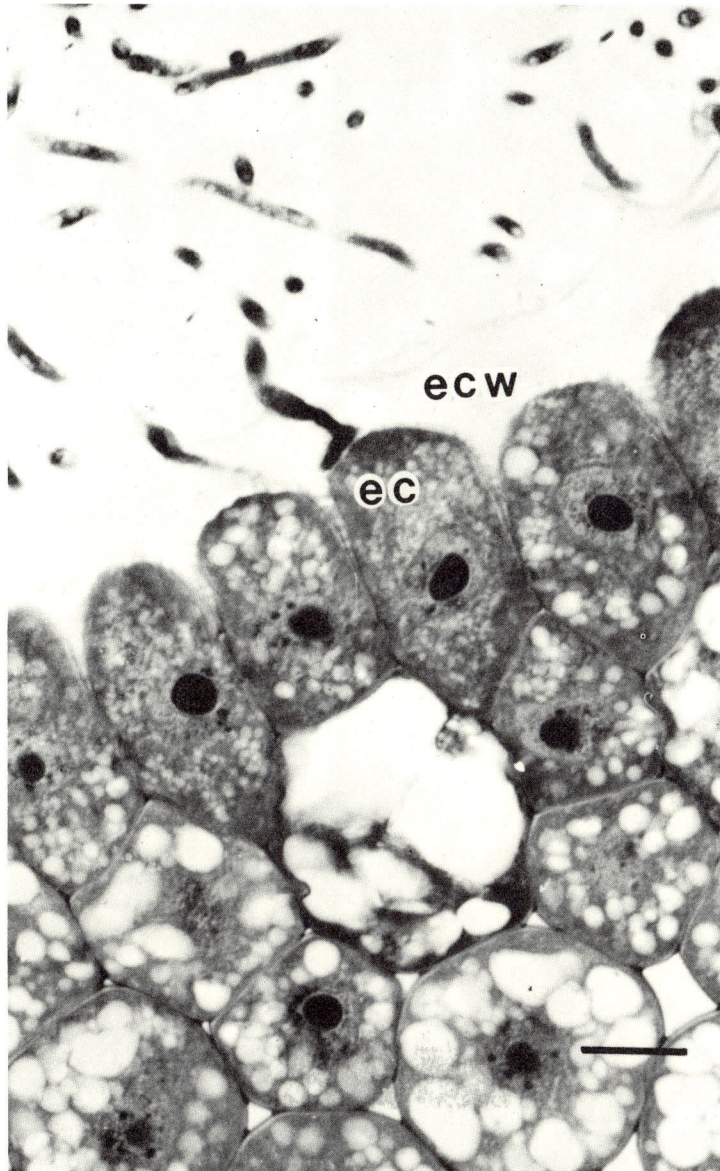

FIG. 1. *Infection of* Zea mays *roots by* Phytophthora cinnamomi. *Light micrographs of root surface 4 hr after infection with low concentrations of zoospores. The epidermal cells appear unaffected by the invasion, and growth is usually arrested in the epidermal cell layer. ec epidermal cell; ew epidermal cell wall. x 970.*

Cell Wall Components

The current concept of a primary cell wall is that it is a gel-like matrix of polysaccharides and proteins in which cellulose microfibrils are embedded. The nature and proportions of the different components change during the life of the cell and in response to a number of environmental, physiological and chemical stimuli, as well as to infection and wounding (Fincher and Stone, 1981; Kato, 1981). Briefly, the four major groups of wall components are:

(i)   Cellulose, to which other wall components including "the high-hydroxyproline containing" glycoprotein, and the non-cellulosic polysaccharide xyloglucan may be non-covalently bound.

(ii) The pectic polymers (Aspinall, 1981), which are essentially polygalacturonic acid linked in 1,4-α-configuration interspersed with 1,2-α-linked rhamnosyl linkages and substituted to a variable extent with glycosyl residues (particularly arabinogalactan residues). The galacturonic acid may be methylated. The degrees of methylation and substitution control the number of regions which are available to interact with other wall polysaccharides to form stable "junction zones."

(iii) Other non-cellulosic polysaccharides ("hemicelluloses") (Aspinall, 1981). These are a variable group of polysaccharides and consist of a glycan backbone, side-branched with other glycosyl polymers or residues. The backbone has a variable degree of substitution. The nature and frequency of these substituents control the regions of the polysaccharides which are available for interactions with other wall components.

(iv) Proteins, including structural glycoproteins and enzymes (Lamport and Catt, 1981), such as hydrolases, which may alter the chain length or degree of substitution of individual wall polymers and hence alter their capacity for interaction.

This simple classification rather obscures the great complexity of composition of some of the wall components, especially the "pectic polysaccharides" and the non-cellulosic polysaccharides. For example, a structure for the pectic polysaccharide is:

$$\rightarrow 4)\text{Gal}p\text{A}\xrightarrow{\alpha}(1\xrightarrow{\alpha}[\ \rightarrow 4)-\text{Gal}p\text{A}-(1\xrightarrow{\alpha}]_n\ \rightarrow 4)-\text{Gal}p\text{A}-(1\ \xrightarrow{\alpha}\ 2)-\text{Rha}p-(1\underline{\quad\quad}$$

```
3                                    3          4
↑                                    ↑          ↑
R                                    R          R'
```

$$R = \text{Xyl}p-(1\xrightarrow{\beta},$$

$$\text{Gal}p-(1\xrightarrow{\beta}2)-\text{Xyl}p-(1\underline{\quad\quad},$$

$$\text{Fuc}p-(1\xrightarrow{\alpha}2)-\text{Xyl}p-(1\underline{\quad\quad},\ \text{or}$$

$$\text{Api}f-(1\longrightarrow 3)-\text{Api}f-(1\underline{\quad\quad}$$

144

```
R or R' = Araf-(1———,

          [Araf]ₘ———, or

          [Araf]ₓ-[Galp]ᵧ———
```

GalpA residues as methyl esters

O-Acetyl groups (location not known)   (after Aspinall, 1981)

FIG. 2 A structure for part of a pectic polysaccharide (adapted from Aspinall, 1981).

Where the diverse nature of the R groups reflects the potential complexity of the molecules.

Relatively few types of cell walls have been analysed. Although in a few studies considerable effort has been invested in obtaining homogeneous cell wall preparations, analyses are more often performed on preparations obtained directly from whole tissues as there are no reliable, standard procedures for isolation of walls of particular cell types (Harris, 1982). Analysis of components within a particular wall preparation is fraught with difficulties and uncertainties. Solubilization and isolation of individual components without degradation are often difficult, although rapid fractionation procedures by lectin affinity and gell exclusion chromatography should be useful. Having obtained a homogeneous preparation of a component from a particular cell type, there are still considerable technical difficulties to obtaining a complete sequence of monosaccharides and their linkages (Valent *et al.*, 1980).

Localization of Individual Components Within Particular Cell Walls

Localization of individual components within the cell wall may be crucial to understanding their function in cell-cell interactions. One approach to defining their organization is to selectively degrade particular components with highly specific enzymes. For example, Roland and Vian (1981) treated sections of mung bean hypocotyls with purified polygalacturonase, and visualized the remaining polysaccharides using the periodic acid-thiosemicarbohydrazide-silver proteinate stain for polysaccharides. Two regions, the middle lamella, and the inner zone close to the cytoplasm, were resistant to enzymic degradation. Possibly this pattern of degradation reflects different degrees of esterification of particular pectic polysaccharides, which specify their vulnerability to enzymic hydrolysis.

A second approach is to examine a particular cell wall for its capacity to bind with specific probes such as antibodies or lectins. For example, in the zone of elongation of *Zea mays* roots, the epidermal cells bind *Ulex* lectin and soybean agglutinin, but not ConA. This pattern of lectin binding reflects a high concentration of accessible fucose and galactose residues in the epidermal cell walls. Binding of both these lectins is notably absent from the walls of the cortical cells. Examination of micrographs also shows that lectin binding is not evenly distributed throughout the epidermal cell wall: only the outer wall and part of the lateral wall bind the lectin strongly (Hinch *et al.*, 1983). Both these approaches give valuable but incomplete information on the distribution of particular polysaccharides in particular cell walls (Fig. 3).

Interaction of Wall Components to Give the Gel Matrix

Pectic polymers and other non-cellulosic polysaccharides interact to form a gel matrix (Kato, 1981) in which the cellulose microfibrils are embedded. The polysaccharides are not ordered in regular arrays, but interact non-covalently at particular regions of individual polymers ("junction zones") to form three dimensional networks. The matrix is an effective diffusion barrier; the size of molecules excluded by callus cells and root hairs is estimated at MW 17,000 for a globular protein, corresponding to an effective "pore" size of 5nm (Carpita *et al.*, 1979). The cell walls of pollen tubes and stigmatic papillae have apparent "pore" sizes of the same order (Hoggart and Clarke, 1983). The gel properties of the cell wall, and hence the exclusion limit to the size of diffusing molecules, can be influenced dramatically by non-covalent interactions of the wall components. A model for this type of interaction is that between two polysaccharides, galactomannan and agarose (Morris, 1979). If solutions of galactomannan and agarose, both at non-gelling concentrations ($<0.1\%$) are mixed, firm rubbery gells are formed. Galactomannans with a relatively low proportion of galactose residues are more effective in the gelling interaction (Dea, 1979). The galactomannan has a 1,4-β-mannosyl backbone with 1,6-β-linked galactose side-branches arranged so that there are "hairy" (substituted) regions interspersed with "smooth" (unsubstituted) regions. Where the smooth regions are about 15-20 residues long, interaction with the regular helical regions of the agarose chains ("junction zones") can occur. The ability of the interacting polymers to form "junction zones" determines the properties of the resulting gel. Thus, properties such as the degree of substitution of either of the interacting polymers will determine the nature and number of potential "junction zones" and hence their potential for gel formation. "Junction zone" formation is, for example, prevented by the presence of residues incompatible with the required ordered conformation and can be competitively inhibited by the presence of polysaccharide chains long enough to form one cooperative junction, but not two. These short chains can occupy binding sites on intact polymer chains without contributing to intermolecular cross linking, and thus disrupt network formation (Morris, 1979). These sorts of effects within the cell walls may quite dramatically influence the porosity of the cell walls and hence the transmission of extracel-

146

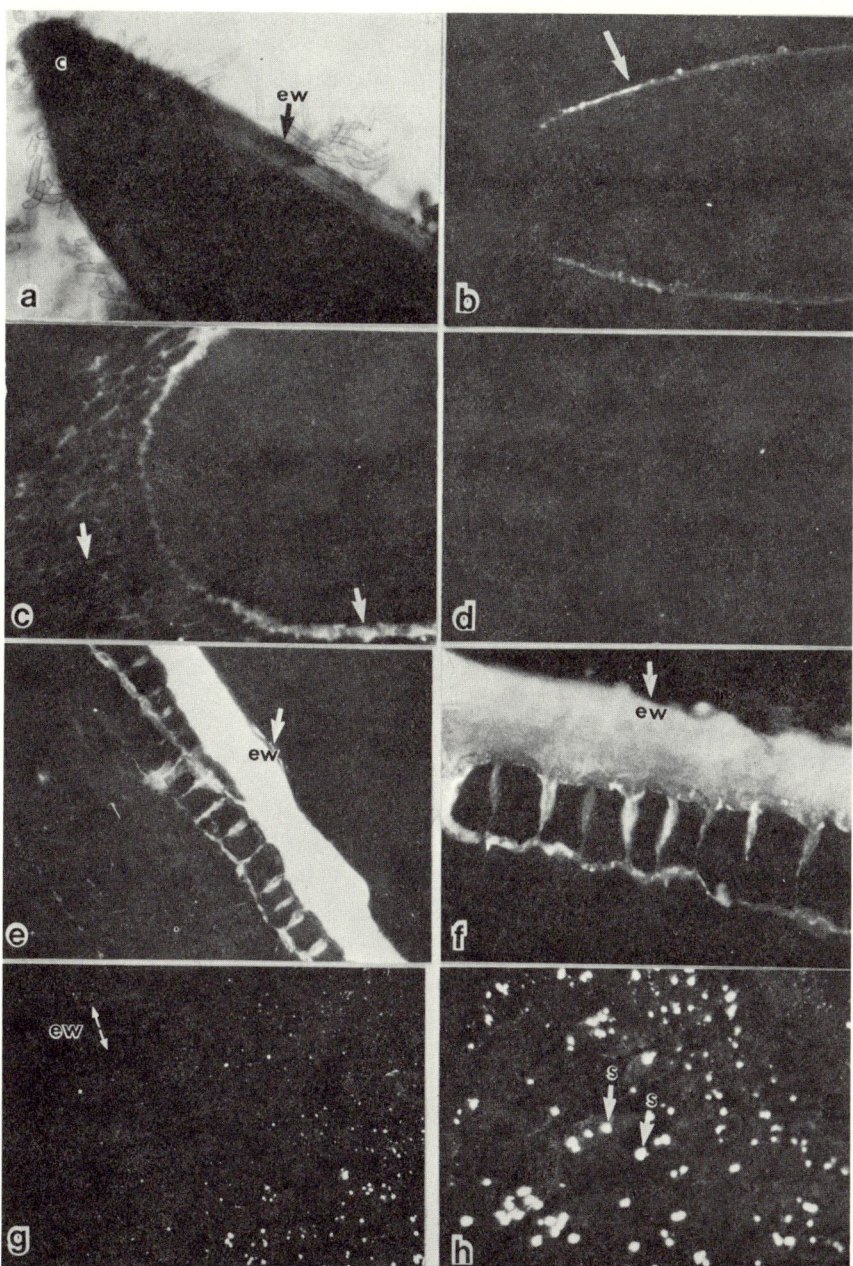

FIG. 3. Lectin binding to Zea mays roots. a. Whole mount of root stained with toluidine blue. c cap; ew epidermal cell wall. x 50. b. Section of root embedded in JB-4 resin, stained with FITC-Ulex lectin. There is strong fluorescence in the zone of elongation (arrow) and weaker fluorescence in the cap region. x 50. c. As for (b), showing more detail of the slime staining in the cap region. Essentially

the same distribution of fluorescence was obtained with both FITC-Ulex lectin and FITC-soybean agglutinin. ew epidermal cell wall. x 100. d. No fluorescence was detected in control sections, in which the lectin was preincubated with the complementary sugar (L-fucose for Ulex lectin; D-galactose for soybean agglutinin). e. Detail of region marked with arrow in (b) stained with FITC-Ulex lectin. The fluorescence is associated with the thick outer epidermal cell wall (ew) as well as the lateral walls. x 250. f. Detail of region corresponding to that marked with arrow in (b) stained with FITC-soybean agglutinin. The distribution of staining is similar to that shown in (e). ew epidermal cell wall. x 625. g. Detail of region corresponding to that marked with arrow in (b), stained with FITC-Concanavalin A. There is no staining in the epidermal cell wall, but scattered staining of intracellular inclusions (probably starch grains) is apparent. x 250. h. Section including part of the cap, stained with FITC-Concanavalin A showing the staining of intracellular inclusions (s) in more detail. x 625.

lular signals. In addition to the effect on "junction zone" formation, the nature and number of substituents will determine the suitability of a polysaccharide as a substrate for particular hydrolytic enzymes.

Experimental Evidence for the Involvement of Cell Walls in Cell-Cell Recognition

Elicitors of phytoalexin production—Both fungal and host cell wall polymers are able to induce production of phytoalexins in some systems. Many different molecules of fungal origin such as polysaccharides, proteins, glycoproteins and lipids have been implicated as elicitors of phytoalexins. One intensively studied elicitor is the β-glucan secreted into the culture medium of *Phytophthora megasperma* var. *sojae*. The elicitor is assumed to originate in the fungal cell wall. A tentative structure for the elicitor is a backbone of glucosyl residues joined in a 1,6-β-linkage, some of which are substituted at C3 by glucosyl residues (Albersheim and Valent, 1978) (Fig. 4). Thus, a precise shape is required for activity. A detailed account of the nature of other fungal elicitors is given by West (1981).

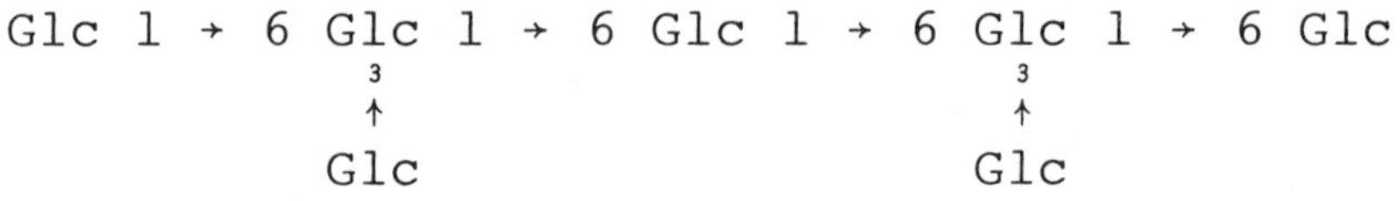

$$\text{Glc } 1 \to 6 \text{ Glc } 1 \to 6 \text{ Glc } 1 \to 6 \text{ Glc } 1 \to 6 \text{ Glc}$$
$$33$$
$$\uparrow\uparrow$$
$$\text{Glc}\text{Glc}$$

FIG. 4. *A structure for the β-glucan elicitor from* Phytophthora megasperma. *(from Albersheim and Valent, 1978).*

148

In addition to this fungal elicitor, an "endogenous" elicitor originating from cell walls of soybeans and other plants has been described. It is apparently a fragment of the "pectic polysaccharide" containing 10-15 galacturonic acid residues (Hahn et al., 1981). Recently, West and co-workers (Lee and West, 1981; Bruce and West, 1982) have shown that an α-1, 4-endopolygalacturonase from culture filtrates of *Rhizopus stolonifer* elicits formation of the phytoalexin casbene in castor bean seedlings. There is convincing evidence that the heat-labile enzyme functions as an elicitor by releasing a heat-stable fragment from the pectic polysaccharide fraction of cell walls from castor bean seedlings. That is, the pectic fragment is an intermediate in the elicitation of casbene initiated by the fungal polygalacturonase.

Differential agglutination of pathogenic microorganisms by cell wall fractions—Kojima and co-workers (1982) have demonstrated that a cell wall fraction has differential spore agglutinating activities. They isolated host specific strains of *Ceratocystis fimbriata* from black rot lesions of sweet potato, coffee, prune, cocao oak and taro. Each of the strains show a strict host specificity (e.g., the sweet potato strain is pathogenic only to sweet potatoes not to the other hosts). A cell wall fraction from sweet potato, containing 53% galacturonic acid, caused agglutination of ungerminated spores of all strains. However there was differential agglutination of germinated spores, with all strains except those from sweet potato and almond being agglutinated. Neither the structure of the active wall fragment nor basis of the agglutination are known, but these observations indicate the potential of cell wall components for immobilizing pathogens as well as eliciting defense responses. Another role of the cell wall is encapsulation of some non-pathogenic, but not pathogenic, bacteria within the intercellular spaces of the host tissue (Whately and Sequiera, 1980).

Proteinase inhibitors accumulated in plants in response to insect attack and wounding—One response of plants to wounding (e.g., insect attack, mechanical damage) is the accumulation of proteinase inhibitors (Bishop et al., 1981). The release of these factors is initiated by fragments of cell wall termed PIIF (proteinase inhibitor inducing factor). The active wall component is a "pectic polysaccharide"—molecular weight 200,000. The backbone, like that of the "endogeneous elicitor," is a polymer of galacturonic acid, interspersed with rhamnosyl residues and side branched with a variety of chains containing arabinosyl and galactosyl residues (Albersheim et al., 1981). The minimal structure compatible with activity is not known.

Adhesion mediated by cell wall components—Cell surface components may be involved in the initial stages of adhesion of microorganisms to plants. For example, attachment of the crown gall bacteria *Agrobacterium tumafaciens* to host cell walls during tumor initiation is inhibited by pretreatment of leaves with cell wall preparations, particularly galacturonic acid containing pectic fractions (Lippincott and Lippincott, 1977; Lippincott et al., 1977). Some wall preparations from monocots do not inhibit tumor production. After treatment with pectin methylesterase, however, these wall preparations become active inhibitors of tumor production (Whately and Sequiera, 1981). Another example is the involvement of root surface mucilage in the adhesion of zoospores. In the interaction of *Phytophthora cinnamomi* with a non-host, *Zea mays,* attachment occurs preferentially in the zone of elongation, to the extracellular mucilage which overlays the thick outer cell wall of the epidermal cells. The composition of this layer is not known, but may be related to the material which is secreted by roots grown in water culture. This "secreted slime" would also contain material originating from the root cap. The monosaccharide composition (%) of this material is known (Gal 21.4, Fuc 19.7, Man 1.0, Ara 10.8, Glc 23.2, Xyl 8.3, Uronic Acid 15.6) (Green and Northcote, 1978), but there is no information available on the sequence of sugars within individual components of the slime, except that a proportion of the fucose is present as terminal non-reducing groups. Modifying the surface of the root either by masking fucosyl residues (by treating whole roots with the *Ulex* lectin) or removing the terminal fucose (with α-L-fucosidase) causes a dramatic inhibition of zoospore adhesion. In contrast, treatment with ConA or a galactosyl binding lectin, Tridacnin, does not alter the adhesive capacity. This implies that the sequence of sugars of root slime is involved in the adhesion of zoospores (Hinch and Clarke, 1980).

Lectin-glycoconjugate interactions—Lectins are involved in several cell recognition systems in both animals (Ashford and Harford, 1981) and lower plants (Barondes, 1981; Evans et al., 1982). Since lectins are widely distributed in higher plants (Kauss, 1981), they are prime candidates for involvement in cell recognition systems in higher plants. Although there are many cases in which lectins apparently do play a role, in no case has both the lectin and its glycoconjugate receptor been isolated, and the cytoplasmic response shown to depend on their interaction. Some of the most convincing evidence is in the *Rhizobium*-legume interaction, but unequivocal evidence for lectin involvement at the root hair-*Rhizobium* contact region is not yet available (Bauer, 1981; Schmidt and Bohlool, 1981). A major gap in the information available concerns the specificity and nature of lectins which occur in situations apart from seed. Experimental work aimed at establishing the role of lectins in the root-microorganism interactions has often been performed with seed lectins, which may or may not be identical with those assumed to be present at the root surface (Schmidt and Bohlool, 1981). Lectins are associated with certain cell wall preparations (e.g., hypocotyls of mung bean, *Vigna radiata*) and can be solubilized with detergent (Haass et al., 1981). The significance of this and of the copurification of wall-lectin and α-galactosidase (Hankins *et al.*, 1980) is not clear. Another curious and incompletely understood observation is the structural relationship between the lectins of the Solanaceae and cell wall glycoprotein (for review see Lamport and Catt. 1982; Fincher *et al.*, 1983).

Cell wall modifications as a response to infection—Apart from being involved in the adhesion of microorganisms and the defense of a plant against infection, the cell walls of host plants may themselves be modified in response to infection. For example, wall appositions are commonly found in the regions of contact between a plant cell and a fungal invader (Aist, 1976; Heath, 1980; Bell, 1981). These appositions contain a $1,3-\beta$-glucan but are often quite heterogenous and may contain silicon and be suberized or lignified. One of the reasons that the development of these wall appositions has been studied intensively is that they give a characteristic fluorescence when stained with decolourized aniline blue. This procedure, taken to be diagnostic of the presence of "callose", is thought to be due to the interaction of the fluorochrome present as an impurity in the commercial aniline blue, with a $1,3-\beta$-glucan or a $1,3;1,4-\beta$-glucan (Smith and McCully, 1978; Hinch and Clarke, 1982). Another common wall modification is lignification in the vicinity of a fungal infection (Bell, 1981). Wall components such as glycoproteins (Esquerré-Tugaye *et al.*, 1979) and soluble arabinogalactans (Hinch and Clarke, 1980b) may also increase in response to certain infections.

The precise relationship of these responses to the final outcome of the interaction between host and pathogen is not known. Thus, the issue of whether wall components are laid down and/or modified for the purpose of impeding fungal growth or whether they are merely a response to wounding (either chemical or mechanical) caused by the pathogen is not established.

PLANT PLASMA MEMBRANES IN CELL-CELL INTERACTIONS

Apart from the involvement of the wall, transfer of extracellular signals to the cytoplasm must involve the plasma membrane. Direct information regarding the role of the plasma membrane in plant cell-cell recognition is restricted to systems in which at least one of the interacting cells is without walls, e.g. *Chlamydomonas* mating (Van den Ende, 1982). Even in these cases, in which the membrane is accessible to probes, our knowledge of the nature of membrane components is poor compared with that of membrane components of animal cells. Not only are some components of several types of animal plasma membranes defined, but there have been rapid advances in the understanding of the mechanisms by which receptors and extracellular signals or ligands interact. In many cell types, there is localized microaggregation of the receptor-ligand complex to form coated pits which then undergo endocytosis. The internalized receptors themselves may participate in subsequent cytoplasmic events, including lysosomal degradation and plasma membrane recycling (Ashwell and Harford, 1982).

Most work on plasma membranes of higher plants has utilized protoplasts, prepared by enzymic degradation of the cell wall (e.g., of callus or leaf mesophyll cells). The end point at which the wall is completely degraded, without altering the membrane, is usually monitored microscopically using, for example, the disappearance of Calcofluor staining. This is unsatisfactory, however, since the end point required is complete removal of the wall components without destruction of putative membrane components, such as glycoconjugates (i.e., a molecular end point). No alternative indicator has been devised, and most of the limited information available has been obtained using protoplasts, although some electrophoretic studies on enriched plasma membrane fractions show the spectrum of molecular species which may be present in a plasma membrane (Booz and Travis, 1980).

Three classes of molecules have been identified on protoplast surfaces:
1. Glycoconjugates. The existence of this class of molecule is implied from the ability of various lectins to specifically agglutinate isolated protoplasts (Larkin, 1981). A logical extension of these observations are the studies by Berkowitz and Travis (1982) on the binding of ConA to plasma membrane enriched fractions prepared from developing soybean roots. The membrane fractions isolated from meristematic tissue bound more ConA than fractions from mature tissue, possibly reflecting either a change in the structure of oligosaccharide side chains of membrane glycoconjugates or a decrease in the concentration of a particular ConA receptor during development. These observations may parallel the finding that in animal cells there are changes in membrane glycoproteins and glycolipids which can be detected immunologically and which are due to addition or deletion of monosaccharide residues or sequences (Feizi, 1981). The function of these antigens is not established, but they may be involved in the control of cooperative interactions between cells.
2. Arabinogalactan-proteins (AGP's). These proteoglycans are present in many plant cell secretions and there is also good indirect evidence that they are associated with the plasma membrane (review Fincher et al., 1983). The carbohydrate moiety of the soluble AGP's characteristically has a branched β-galactopyranose framework with predominantly 1,3-linkages and varying amounts of 1,6-linkages. Analyses are compatible with a structure having a linear backbone with short side branches (Fig. 5). The branches may contain predominantly galactosyl and arabinosyl residues as well as a variety of other

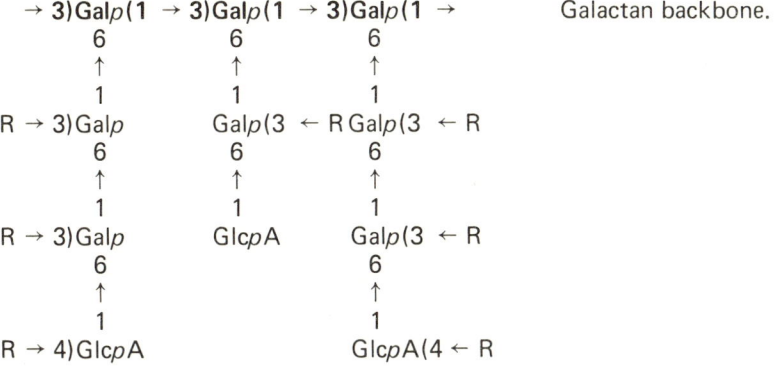

$$\rightarrow 3)\text{Gal}p(1 \rightarrow 3)\text{Gal}p(1 \rightarrow 3)\text{Gal}p(1 \rightarrow \qquad \text{Galactan backbone.}$$

```
→ 3)Galp(1 → 3)Galp(1 → 3)Galp(1 →        Galactan backbone.
      6             6             6
      ↑             ↑             ↑
      1             1             1
R → 3)Galp       Galp(3 ← R Galp(3 ← R
      6             6             6
      ↑             ↑             ↑
      1             1             1
R → 3)Galp       GlcpA         Galp(3 ← R
      6                           6
      ↑                           ↑
      1                           1
R → 4)GlcpA                     GlcpA(4 ← R
```

R = Rhap(1, Araf(1, Galp(1 → 3)Araf(1, Araf(1 → 3)Araf(1.

(adapted from Fincher *et al.* 1983)

FIG. 5. *A structure for the carbohydrate moiety of a soluble arabinogalactan protein. (adapted from Fincher et al., 1983).*

monosaccharides. It is possible that the outer chains have an informational role, in that the nature and organization of these side branch saccharides may be involved in the expression of identity of individual tissues or cell types. The presence of free galactosyl residues on the surface of protoplasts of *Lolium multiflorum* has been used to design a method for isolation of plasma membranes, based on the binding of a galactosyl-specific IgG to the protoplast surface (Schibeci *et al.*, 1982). This type of approach will be most valuable in ultimately defining the nature and function of plasma membrane components.

3. Antigens. Certain plant antigens are associated with the plasma membrane. For example, an IgG cut of antiserum raised to leaf callus will bind to leaf protoplasts but not to callus cells (Raff *et al.*, 1980). This indicates that when a callus homogenate is injected into a rabbit, the rabbit immune system perceives and responds to plasma membrane-associated molecules in preference to wall-associated molecules. The fact that these molecules are antigenic means only that they elicit IgG production in an experimental animal and has no implications for immunity in the plant. However, not only are there antigenic molecules present at plant plasma membranes, but there are also antigens apparently characteristic of particular tissue types (Raff *et al.*, 1980). There is no information regarding either the structure or function of these molecules, but it is possible that they are similar to their animal counterparts which have been intensively studied. In animals they may act as "area code" markers which deter-

mine the differentiation pathway of a particular cell and also act as specific surface receptors of extracellular signals (Hood *et al.*, 1977). The approach of using antibodies, especially monoclonal antibodies, to identify the structure and function of plant membrane components has great, and as yet, unexploited potential.

POSSIBLE MECHANISMS FOR PLANT CELL-CELL INTERACTIONS

As no single cell-cell interaction involving a higher plant is known in detail, any mechanism proposed must be regarded as speculative. All extracellular signals must be transmitted across the wall in some way to reach the plasma membrane and eventually elicit the cytoplasmic response. On the basis of our knowledge of the cell wall structure, and of the fragments of information available for a few cell recognition systems, there are several possible mechanisms (Fig. 6):

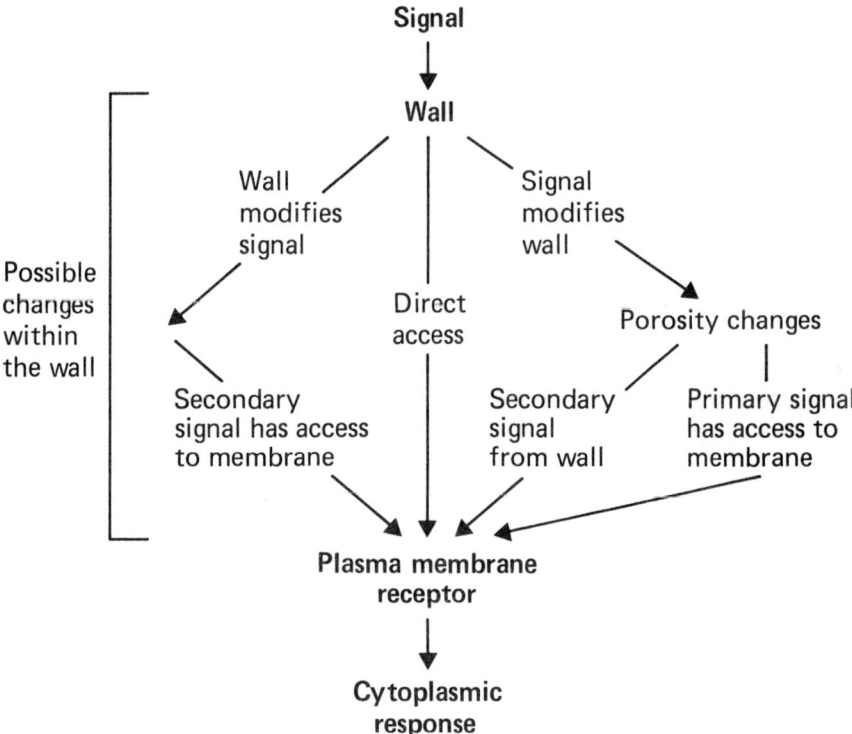

FIG. 6. *Schematic representation of possible interactions between the cell wall and an external signal during cell-cell recognition in plants.*

Direct Access of Signal to Membrane

In this case the wall would act merely as a molecular or ionic sieve allowing access of molecules of particular size or charge to complementary receptors.

The Signal Modifies the Wall

The primary extracellular signal may be an enzyme which modifies a particular component (e.g., release of a fragment of wall, such as a protein or polysaccharide). This modification of a particular component may induce changes in the wall porosity by modifying the number of "junction zones" and hence the gel properties of the wall. Alternatively, by changing the wall porosity, the primary signal (enzyme) may gain access to the membrane from which it was initially excluded because of its size. In addition to inducing changes in porosity, the wall fragment itself may act as a secondary signal, and interact directly with a plasma membrane receptor.

The Wall Modifies the Signal

It is also possible that the wall components themselves (e.g., enzymes, lectins) may modify the primary signal and in so doing generate a secondary signal which has access to the putative membrane receptor.

Our current information is insufficient to exclude any of these possibilities, or to assign unequivocally any mechanism for a particular cell-cell interaction. However, evidence from the action of pectic polymers and polygalacturonases in eliciting phytoalexin production implicates a mechanism involving a modification of the wall by the original signal. It is also relevant that in the yeast mating system, agglutination of compatible mating types is mediated by interaction of wall-associated components on the interacting cells (Pierce and Ballou, 1982). Unfortunately the mechanism by which this interaction signals wall and membrane fusion is not known.

CONCLUSIONS AND SUMMARY

Cell-cell recognition in plants is poorly understood at the molecular level compared with cell recognition in animal systems. One reason for this lack of knowledge is the relatively little experimental effort in the study of plant systems compared with that in the medically important cell-cell recognition systems involved in infection, control of the immune response, development of cancer and tissue transplantation. The principles of cell recognition derived from animal systems cannot be applied directly to plant systems, since plant cells possess a cell wall which overlays the plasma membrane. All extracellular signals must traverse the wall to reach the plasma membrane. The wall is a dynamic gel-like matrix

155

in which cellulose microfibrils are embedded. The composition of the wall matrix (like that of animal membranes and extracellular matrix components) changes during the life of the cell. Evidence (mainly from work on the molecular basis of plant pathogenesis) indicates that the cell wall may not merely present a barrier to diffusion of extracellular signals to putative membrane receptors, but that secondary signals (e.g., fragments of wall polymers) may be generated by the action of a primary (enzymic) signal. Isolation and analysis of the cell wall polymers, particularly the "pectic polymers", poses considerable technical difficulties, but it will be essential to define the structures of the wall components and their fragments as a basis for understanding their role in cell-cell recognition. The nature of the putative receptors for the fragments of wall polymers (or other signals) at the plasma membrane is essentially unexplored. Indeed, there is little information available concerning the nature of plant plasma membrane components or the function of particular components. The approach of using antibodies and lectins to establish the tissue distribution of membrane components, to isolate particular components, and to probe their function has been used with marked success in animal systems, and similar approaches could be applied to protoplasts to obtain similar information for plant plasma membranes.

There are urgent practical incentives for establishing the molecular basis of plant cell recognition. Recognition phenomena are fundamental to understanding host-pathogen interactions and hence the molecular basis of plant disease. Similar phenomena apparently control pollen-stigma interactions and, hence, plant fertilization. Thus, understanding cell-cell interactions may lead to procedures for minimising crop losses through disease and for breaching incompatibility barriers to allow breeding of more productive hybrids. Our ultimate understanding will require a thorough understanding of the structure and function of both the plant plasma membrane and the cell wall.

ACKNOWLEDGEMENTS

The authors thank Ms. Ann Pottage for drawing Fig. 6, Ms. Joan Mallet and Ms. Ingrid Bonig for assistance in preparing Fig. 1 and Fig. 3, and Dr. Jillian Hinch for providing the material used in Fig. 1 and Fig. 3. We would also like to acknowledge the help we have had from our colleagues and graduate students during discussions which contributed to the ideas formulated in the chapter. Finally, we are most grateful to Dr. Jillian Hinch and Professor B. A. Stone for their critical appraisal of the manuscript and to Ms. G. Loy for expert and patient editorial assistance.

LITERATURE CITED

Aist, J. R. (1976). Papillae and related wound plugs of plant cells. *Ann. Rev. Phytopathol. 14*, 145-163.

Albersheim, P., McNeil, M., Darvill, A. G., Valent, B. S., Hahn, M. G., Robertson, B. K. and Aman, P. (1981). Structure and function of complex carbohydrates active in regulating the interactions of plants and their pests. In: *Recent Advances in Phytochemistry* Vol. 15. The phytochemistry of cell recognition and cell surface interactions. F. A. Loewus and C. A. Ryan (eds.). Plenum Press, New York and London. pp. 37-58.

Albersheim, P. and Valent, B. S. (1978). Host-pathogen interactions in plants. Plants when exposed to oligosaccharides of fungal origin, defend themselves by accumulating antibodies. *J. Cell Biol. 78*, 627-643.

Ashwell, G. and Harford, J. (1982). Carbohydrate-specific receptors of the liver. *Ann. Rev. Biochem. 51*, 531-554.

Aspinall, G. O. (1981). Constitution of plant cell wall polysaccharides. In: *Encyclopedia of Plant Physiology* New Series Vol. 13B. Plant Carbohydrates II, Extracellular Carbohydrates. W. Tanner and F. A. Loewus (eds.). Springer-Verlag, Berlin and New York. pp. 3-8.

Barondes, S. H. (1981). Lectins: Their multiple endogenous cellular functions. *Ann. Rev. Biochem. 50*, 207-232.

Bauer, W. D. (1981). Infection of legumes by Rhizobia. *Ann. Rev. Plant Physiol. 32*, 407-449.

Bell, A. (1981). Biochemical mechanisms of disease resistance. *Ann. Rev. Plant Physiol. 32*, 21-81.

Berkowitz, R. L. and Travis, R. L. (1982). A comparative evaluation of the level of Concanavalin A binding by enriched plasma membrane fractions from developing soybean roots. *Plant Physiol. 69*, 379-384.

Bishop, P. D., Makus, D. J., Pearce, G. and Ryan, C. A. (1981). Proteinase inhibitor-inducing factor activity in tomato leaves resides in oligosaccharides enzymically released from cell walls. *Proc. Natl. Acad. Sci. U.S.A. 78*, 3536-3540.

Booz, M. L. and Travis, R. L. (1980). Electrophoretic comparison of polypeptides from enriched plasma membrane fractions from developing soybean roots. *Plant Physiol. 66*, 1037-1043.

Bruce, R. J. and West, C. A. (1982). Elicitation of casbene synthetase activity in castor bean. The role of pectic fragments of the plant cell wall in elicitation by a fungal endopolygalacturonase. *Plant Physiol. 69*, 1181-1188.

Carpita, N., Dubularse, D., Montezinos, D. and Delmer, D. P. (1979). Determination of the pore size of cell walls of living plant cells. *Science 205*, 1144-1147.

Chorney, M. J. and Cheng, T. C. (1980). Discrimination of self and non-self in invertebrates. In: J. J. Marchalonis and N. Cohen (eds.) *Contemp. Topics in Immunobiol. 9*, 37-54.

Clarke, A. E. and Gleeson, P. A. (1981). Molecular aspects of recognition and response in the pollen-stigma interaction. In: *Recent Advances in Phytochemistry* Vol. 15 The Phytochemistry of Cell Recognition and Cell Surface Interactions. F. A. Loewus and C. A. Ryan (eds.) Plenum Press, New York and London. pp. 161-211.

Clarke, A. E. and Hoggart, R. M. (1982). The use of lectins in the study of glycoproteins. In: *Antibody as a Tool: The Applications of Immunochemistry*. J. J. Marchalonis and G. W. Warr (eds.). John Wiley & Sons, New York. pp. 347-401.

Clarke, A. E. and Knox, R. B. (1978). Cell recognition in flowering plants. *Q. Rev. Biol. 53*, 3-28.

Clarke, A. E. and Knox, R. B. (1979). Plants and immunity. *Develop. and Comp. Immunol. 3*, 557-589.

Curtis, A. S. G. ed. (1978). Cell-Cell Recognition. *Symp. Soc. Exp. Biol.* Cambridge University Press.

Dea. I. C. M. (1979). Interactions of ordered polysaccharide structures—synergism and freeze-thaw phenomena. In: *Polysaccharides in Food.* J. M. V. Blanshard and J. R. Mitchell (eds.) Butterworths, London and Boston. pp. 220-247.

Deverall, B. J. (1977). Defence mechanisms of plants. *Cambridge Monographs in Experimental Biology No. 19.* Cambridge University Press.

Esquerré-Tugayé, M. T., Lafitte, C., Mazau, D., Toppan, A., Touzé, A. (1979). Cell surfaces in plant-microorganism interactions. II. Evidence for the accumulation of hydroxyporline-rich glycoproteins in the cell wall of disease plants as a defence mechanism. *Plant Physiol. 64*, 320-326.

Evans, L. V., Callow, J. A. and Callow, M. E. (1982). The biology and biochemistry of reproduction and early development in *Fucus.* Progress in Phycological Research *1*, 68-110.

Feizi, T. (1981). Carbohydrate differentiation antigens. *TIBS - December 1981*, 333-335.

Fincher, G. B. and Stone, B. A. (1981). Metabolism of noncellulosic polysaccharides. In: *Encyclopedia of Plant Physiology* New Series, Vol. 13B. Plant Carbohydrates II, Extracellular Carbohydrates. W. Tanner and F. A. Loewus (eds.). Springer-Verlag, Berlin and New York. pp. 68-132.

Fincher, G. B., Stone, B. A. and Clarke, A. E. (1983). Arabinogalactan-proteins: Structure, biosynthesis and function. *Ann. Rev. Plant Physiol. 34*, 47-70.

Green, J. R. and Northcote, D. H. (1978). The structure and function of glycoproteins synthesized during slime-polysaccharide production by membranes of the root cap cells of maize (*Zea mays*). *Biochem. J. 170*, 599-608.

Gunning, B. E. S. (1983). The special features of plant cells In: *Molecular Biology of the Cell.* J. Watson, M. Raff, B. Alberts, D. Bray and K. Roberts (eds.), Garland Inc., New York.

Haass, D., Frey, R., Theisen, M., Kauss, H. (1981). Partial purification of a hemagglutinin associated with the cell walls from hypocotyls of *Vigna radiata. Planta 151*, 490-496.

Hahan, M. G., Darvill, A. G. and Albersheim, P. (1981). Host-pathogen interactions XIX. The endogenous elicitor, a fragment of a plant cell wall polysaccharide that elicits phytoalexin accumulation in soybeans. *Plant Physiol. 68*, 1161-1169.

Hankins, C. N. Kindinger, J. I., Shannon, L. J. (1980). Legume α-galactosidases which have hemagglutinin properties. *Plant Physiol. 65*, 618-622.

Harris, P. J. (1982). Cell Walls. In: *Isolation of Membranes and Organelles from Plant Cells.* J. L. Hall and A. L. Moore (eds.). Academic Press.

Heath, M. (1980). Reactions of non-suscepts to fungal pathogens. *Ann. Rev. Phytopathol. 18*, 211-236.

Heslop-Harrison, J. (1975). Incompatibility and the pollen-stigma interaction. *Ann. Rev. Plant Physiol. 26*, 403-425.

Hildeman, W. H. (1974). Some new concepts in immunological phylogeny. *Nature 250*, 116-120.

Hinch, J. M., Boenig, I., and Clarke, A. E. (1983). Structure and distribution of secreted root slime components of legumes and corn. *Aust. J. Plant Physiol.* (submitted).

Hinch, J. M. and Clarke, A. E. (1980a). Adhesion of fungal zoospores to root surfaces is mediated by carbohydrate determinants of the root slime. *Physiol. Plant Pathol. 16*, 303-307.

Hinch, J. M. and Clarke, A. E. (1980b). Increases in soluble arabinogalactans and cell wall associated hydroxyproline in lupins infected with *Phytophthora cinnamomi*. *A.G.P. News* Proc. AGP Club, Melbourne.

Hinch, J. M. and Clarke, A. E. (1982). Callose formation in *Zea mays* as a response to infection with *Phytophthora cinnamomi*. *Physiol. Plant Path.* (in press).

Hinch, J. M. and Weste, G. M. (1979). Behaviour of *Phytophthora cinnamomi* zoospores on roots of Australian forest species. *Aust. J. Bot. 27*, 679-691.

Hood, L., Huang, M. V. and Dreyer, W. J. (1977). The area-code hypothesis: The immune system provides clues to understanding the genetic and molecular basis of cell recognition during development. *J. Supramol. Struct. 7*, 531-539.

Inchley, C. J. (1981). *Immunobiology*. Studies in Biology No. 128. Edward Arnold.

Kato, K. (1981). Ultrastructure of the plant cell wall: Biochemical viewpoint. In: *Encyclopedia of Plant Physiology* New Series, Vol. 13B. Plant Carbohydrates II, Extracellular Carbohydrates. W. Tanner and F. A. Loewus (eds.). Springer-Verlag, Berlin and New York. pp. 29-46.

Kauss, H. (1981). Lectins and their physiological role in slime moulds and in higher plants. In: *Encyclopedia of Plant Physiology* New Series Vol. 13B. Plant Carbohydrates II, Extracellular carbohydrates. W. Tanner and F. A. Loewus (eds.). Springer-Verlag, Berlin. pp. 627-657.

Kojima, M., Kawakita, K. and Uritani, I. (1982). Studies on a factor in sweet potato root which agglutinates spores of *Ceratocystis fimbriata*, Black rot fungus. *Plant Physiol. 69*, 474-478.

Kosuge, T. (1981). Carbohydrates in plant-pathogen interactions. In: *Encyclopedia of Plant Physiology*, New Series Vol. 13B, Plant Carbohydrates II, Extracellular carbohydrates. Eds. W. Tanner and F. A. Loewus. Springer-Verlag, Berlin, Heidelberg, New York. pp. 584-623.

Kuć, J. (1976). Phytoalexins. In: *Physiological Plant Pathology, Encyclopedia of Plant Physiology*, Vol. 4 R. Hertefuss and P. H. Williams (eds.). Springer-Verlag, Berlin and New York. pp. 632-652.

Lamport, D. T. A. and Catt, J. W. (1981). Glycoproteins and enzymes of the cell wall. In: *Encyclopedia of Plant Physiology* New Series Vol. 13B. Plant Carbohydrates II Extracellular Carbohydrates. W. Tanner and F. A. Loewus (eds.). Springer-Verlag, Berlin and New York. pp. 133-165.

Larkin, P. J. (1981). Plant protoplast agglutination and immobilization. In: *Recent Advances in Phytochemistry* Vol. 15. The phytochemistry of cell recognition and cell surface interactions. F. A. Loewus and C. A. Ryan (eds.) Plenum Press, New York and London. pp. 135-160.

Lee, S. C. and West, C. A. (1981). Properties of *Rhizopus stolonifer* polygalacturonase, an elicitor of casbene synthetase activity in castor bean (*Ricinus communis* L.) seedlings. *Plant Physiol. 67*, 640-645.

Lippincott, J. A. and Lippincott, B. B. (1977). Nature and specificity of the bacterium-host attachment in *Agrobacterium* infection. In: *Cell Wall Biochemistry Related to Specificity in Host-Plant Pathogen Interactions*. B. Solheim and J. Raa (eds.). Norway Universitetsforlaget, Oslo. pp. 439-451.

Lippincott, B. B., Whatley, M. H. and Lippincott, J. A. (1977). Tumor induction by *Agrobacterium* involves attachment of the bacterium to a site on the host plant cell wall. *Plant Physiol. 59*, 388-390.

Moore, R. (1981). Graft compatibility and incompatibility in higher plants. *Devel.*

and Comp. Immunol. 5, 377-389.

Morris, E. R. (1979). Polysaccharide structure and conformation in solutions and gels. In: Polysaccharides in Food J. M. V. Blanshard and J. R. Mitchell (eds.). Butterworths, London and Boston. pp. 15-31.

Pierce, M. and Ballou, C. E. (1982). Cell recognition and adhesion in yeast mating. In: Receptors in Plants and Slime Moulds. C. M. Chadwick and D. R. Garrod (eds.). Marcel Dekker Inc., New York.

Raff, J. W., McKenzie, I. F. C. and Clarke. A. E. (1980). Antigenic determinants of Prunus avium are associated with the protoplast surface. Z. Pflanzenphysiol. Bd. 98, 225-234.

Ralton, J. E. and Clarke, A. E. (1982). Comparison of plant and animal recognition systems. In: Cell Receptors and Cell Communication in Invertebrates. A. H. Greenberg (ed.) Marcel Dekker, New York.

Roitt, I. M. (1980). Essential Immunology. 4th Edition. Blackwell Scientific Publications.

Roland, J-C. and Vian B. (1981). Use of purified endopolygalacturonase for a topochemical study of elongating cell walls at the ultrastructural level. J. Cell Sci. 40, 333-343.

Schibeci, A., Fincher, G. B., Stone, B. A. and Wardrop, A. B. (1982). Isolation of plasma membrane from protoplasts of Lolium multiflorum (ryegrass) endosperm cells. Biochem. J. 205 (in press).

Schmidt, E. L. and Bohlool, B. B. (1981). The role of lectins in symbiotic plant-microbe interactions. In: Encyclopedia of Plant Physiology, New Series Vol. 13B, Plant Carbohydrates II, Extracellular carbohydrates. Eds. W. Tanner and F. A. Loewus. Springer-Verlag, Berlin, Heidelberg, New York. pp. 658-677.

Smith, M. M. and McCully, M. E. (1978). A critical evaluation of the specificity of aniline blue induced fluorescence. Protoplasma 95, 229-254.

Turgeon, B. G. and Bauer, W. D. (1982). Early events in the infection of soybean by Rizobium japonicum. Time course and cytology of the initial infection process. Canad. J. Bot. 60, 152-161.

Valent, B. S., Darvill, A. G., McNeil, M., Robertson, B. K. and Albersheim, P. (1980). A general and sensitive chemical method for sequencing the glycosyl residues of complex carbohydrates. Carbohydr. Res. 79, 165-192.

Van den Ende, H. (1981). Sexual interactions in the green alga Chlamydomonas eugamotos. In: Sexual Interactions in Eukaryotic Microbes. Academic Press, New York. pp. 297-318.

Watson, J. D. (1977). Molecular Biology of the Gene. 3rd Edition. W. G. Benjamin, Inc.

West, C. A. (1981). Fungal elicitors of the phytoalexin response in higher plants. Naturwiss. 68, 447-457.

Whatley, M. H. and Sequeira, L. (1981). Bacterial attachment to plant cell walls. In: The Phytochemistry of Cell Recognition and Cell Surface Interactions. F. A. Loewus and C. A. Ryan (eds.). pp. 213-240.

Wiese, L. and Wiese, W. (1978). Sex cell contact in Chlamydomonas, a model for cell recognition. Symp. Soc. Exp. Biol. XXXII, Cell-Cell Recognition. A.S.G. Curtis (ed.) Cambridge University Press.

INDEX

A

C

E

F

G